Mercury and the Making of California

Mining the American West

Duane A. Smith, Robert A. Trennert, and Liping Zhu, editors

Boomtown Blues: Colorado Oil Shale, Andrew Gulliford

From Redstone to Ludlow: John Cleveland Osgood's Struggle against the United Mine Workers of America, F. Darrell Munsell

Gambling on Ore: The Nature of Metal Mining in the United States, 1860–1910, Kent A. Curtis

Hard as the Rock Itself: Place and Identity in the American Mining Town, David Robertson

High Altitude Energy: A History of Fossil Fuels in Colorado, Lee Scamehorn

Industrializing the Rockies: Growth, Competition, and Turmoil in the Coalfields of Colorado and Wyoming, David A. Wolff

The Mechanics of Optimism: Mining Companies, Technology, and the Hot Spring Gold Rush, Montana Territory, 1864–1868, Jeffrey J. Safford

Mercury and the Making of California: Mining, Landscape, and Race, 1840–1890, Andrew Scott Johnston

The Rise of the Silver Queen: Georgetown, Colorado, 1859–1896, Liston E. Leyendecker, Duane A. Smith, and Christine A. Bradley

Santa Rita del Cobre: A Copper Mining Community in New Mexico, Christopher J. Huggard and Terrence M. Humble

Silver Saga: The Story of Caribou, Colorado, Revised Edition, Duane A. Smith

Thomas F. Walsh: Progressive Businessman and Colorado Mining Tycoon, John Stewart

Yellowcake Towns: Uranium Mining Communities in the American West, Michael A. Amundson

Mercury *and the* Making of California

Mining, Landscape, and Race, 1840–1890

ANDREW SCOTT JOHNSTON

UNIVERSITY PRESS OF COLORADO
Boulder

© 2013 by University Press of Colorado

Published by University Press of Colorado
1580 North Logan Street, Suite 660
PMB 39883
Denver, Colorado 80203-1942

All rights reserved
First paperback edition 2026

 The University Press of Colorado is a proud member of
Association of University Presses.

The University Press of Colorado is a cooperative publishing enterprise supported, in part, by Adams State University, Colorado State University, Fort Lewis College, Metropolitan State University of Denver, Regis University, University of Colorado, University of Northern Colorado, Utah State University, and Western State Colorado University.

ISBN: 978-1-60732-242-9 (hardcover)
ISBN: 978-1-60732-462-1 (paperback)
ISBN: 978-1-60732-243-6 (ebook)
https://doi.org/10.5876/9781607322436

Library of Congress Cataloging-in-Publication Data

Johnston, Andrew Scott.
 Mercury and the making of California : mining, landscape, and race, 1840–1890 / Andrew Scott Johnston.
  pages cm
 Includes bibliographical references and index.
 ISBN 978-1-60732-242-9 (hardcover : alk. paper) — ISBN 978-1-60732-243-6 (ebook) — ISBN 978-1-60732-462-1 (pbk : alk. paper)
 1. Mineral industries—California—History—19th century. 2. Mercury ores—California—History—19th century. 3. Landscapes—California—History—19th century. 4. Mining camps—California—History—19th century. 5. California—Economic conditions—19th century. 6. California—Social conditions—19th century. 7. California—History—19th century. I. Title.
 HD9506.U63C356 2013
 338.209794'09034—dc23
                           2013012515

Cover illustrations. Front, top: courtesy, History San Jose; front, bottom: © ktsdesign/Shutterstock.

For Jessica

Contents

Acknowledgments ix

Introduction: California: The Quicksilver State 1

1. Imperialism and California's Quicksilver 21

2. Money and Power in the California Mercury Landscape 57

3. A Geography of Mercury Mining in California 93

4. Race, Space, and Power at New Almaden 137

5. Race, Technology, and Work 193

6. Race, Family, and Camp Life 215

7. Conclusion: The Legacy of the Quicksilver Landscapes of California 245

Bibliography 261

Index 277

Acknowledgments

This book was inspired by work I did as an architect for the Historic American Engineering Record (HAER) of the National Park Service, recording and interpreting the Mariscal Mercury Mine in Big Bend National Park. Eric DeLony, former chief of HAER, has been an enthusiastic supporter of this project. Soon after I finished the Big Bend project, John L. Livermore and Anthony Cerrar took me on a tour of the Oat Hill and Corona Mines in Napa County, and it was on this day that I decided that the mercury mining industry in California would make an excellent research topic. Noel Kirshenbaum arranged this trip, for which I am very grateful. Eleanor Swent was an excellent and enthusiastic traveling companion for bopping around old mercury mine sites, all the way from New Idria to Cloverdale. Her fearlessness led me to often tread where I would not have alone. Her mining oral histories, done through the Regional Oral History Office of the Bancroft Library, added immeasurably to my understandings of mines and mining in California and the West.

A network of colleagues from the fields of mining history and the history of technology has been crucial to the success of this project. Patrick Malone has been a steadfast supporter of this project, and his scholarship on the history of technology and his work ethic are inspirational. I was fortunate to meet Robert Spude and Donald Hardesty through the work in Big Bend. Others I met through the Mining History Association—including Fredric Quivik, Ronald and Susan James, Sally Zanjani, Clark Spence, and Liping Zhu—all took time to speak with me about my work. Duane Smith kindly read the manuscript and provided insightful feedback. At the Sixth International Symposium of the Cultural Heritage in Geosciences, Mining and Metallurgy—held in Idrija, Slovenia—I met Tatjana Dizdarevic, Mirjam Gnezda, Janez Pirc, and Bojan Rezun, each of whom has been of assistance in my research. Luis Jordá Bordehore, whom I also met at this conference, has been very helpful concerning the Almadén Mine.

There is a long list of people whom I met and spoke with over the years, and who added some kernel to information that proved important to my work. These people include Kris Lane, Kenneth Cameron, Henry Glassie, Gerald Weber, Gray Brechin, Christopher Brown, José Peral López, Michael Boulland, Marianne Hurley, Brian Ramos, Rick Fitzgerald, Meg Scantlebury, Thad Van Buren, Sunshine Psota, Arthur Gomez, Bruce Sinclair, Diane Kane, Elizabeth Krase, Elizabeth McKee, and Richard Higgins, for his help with finding the right title.

The Bancroft Library Study Award, the Trent Dames Fellowship at the Huntington Library, and the John Carter Brown Library–Center for New World Comparative Studies all supported my work financially and gave me the opportunity to immerse myself in the collections of these institutions. A Bistline Fellowship provided funding for illustrations and graphic work. A Research Development Fund grant from Xi'an Jiaotong–Liverpool University funded proofreading and indexing. My thanks go to Austin Porter for his graphic work on maps, graphs, and charts. The Institute for Advanced Study in Princeton, New Jersey, provided a stimulating and supportive environment in which to edit and prepare the manuscript. My special thanks also to Joan Scott for her support during this time.

The reference staff and archivists at the Bancroft, Huntington, and John Carter Brown Libraries; the Chemical Heritage Foundation; the California History Room of the California State Library; the National Archive in San Bruno; the Greene Library at Stanford; and the Idrija Municipal Museum were all immensely helpful to this project. Of special note are Dan Lewis at the Huntington Library, Randal Brandt at the Bancroft Library, and Waverly Lowell, formerly at the National Archives and Records Administration. Kitty Monahan, Michael Boulland, and staff at the Almaden Quicksilver Mining Museum were gracious with their time and resources. As well there are the staff and volunteers at the myriad small-town libraries, museums, and historical centers throughout the Coast Ranges of California who aided my research.

Among the faculty at Berkeley I have many people to thank, including Dell Upton, Laurie Wilkie, Allan Pred, Richard Walker, and Galen Cranz. The members of my dissertation writing group, Marie-Alice L'Heureux and Don Choi, gave me excellent advice and support. Other Berkeley students and friends deserving thanks include Tania Martin, Raymond Weschler, Zeynep Kezer, Philip Gruen, and many more.

At the University Press of Colorado I thank Darrin Pratt, Jessica d'Arbonne, Laura Furney, Daniel Pratt, and Beth Svinarich. My thanks also to Sonya Manes for her copyediting. Inky, Dudley, Bob, Maximum, and Photon cats gave me love and companionship and shredded my drafts. My son, Benjamin, is a joy and has an excellent eye for photo interpretation. Jessica Sewell has lived with this project from start to finish, and I thank her for her love and support.

Mercury and the Making of California

California is the Golden State, and has been linked with gold ever since the rush started in 1848. Gold colors our understanding of California and its history; there are elaborate myths of the gold rush emphasizing rugged individualism, democracy, manifest destiny, and cycles of boom and bust. This is the history of the California Dream, a history that excels at stories of innovation, change, dynamism, and reinvention. As a metaphor for California, gold conveys the image of prosperity, youth, and vigor. Being golden also implies opportunity and success, and California has provided opportunity for millions, as well as exploitable resources for tremendous financial success for a few. Writer Carey McWilliams, in his 1949 classic *California: The Great Exception*, argues that the "magic equation" for understanding the exceptionalism of California's development, in relation to the development of the rest of the United States, is "gold-equals-energy." For McWilliams the gold rush and the energizing effect of gold as a catalyst and as an economic pump-primer for the state established the "chain-reaction, explosive, self-generating pattern of development" that has characterized California.[1] McWilliams's insightful ideas guide our historical understandings of the state. However, although gold in many ways is an excellent metaphor for the state, in still other ways it obscures and misrepresents other histories in its blinding glare.

Not all of California history matches the mythical ideals of the gold rush. This book asks what we can learn of the history of California if we move our focus, our metaphor, for California from gold to quicksilver. What if, instead of the Golden State, California were the Quicksilver State? Before the discovery of gold, quicksilver was mined in California—the state was the largest producer of the element in the Western Hemisphere, and the single richest mine of any type in the state was a quicksilver mine. For California, the cultural associations of mercury can be just as fitting as those of gold. Although mercury is associated with abundance and commercial success, it is also associated with eloquence, ingenuity, and thievishness,

Introduction

California: The Quicksilver State

It was, of course, the discovery of gold that got California off to a flying start, and set in motion its chain-reaction, explosive, self-generating pattern of development.
—Carey McWilliams, *California: The Great Exception*, 25

qualities that are often applied to the accumulation of wealth. A mercurial nature implies that one is erratic, volatile, and unstable—all acknowledged characteristics of California's development. Following Carey McWilliams, if we were to invent an equation involving quicksilver for understanding the development of California, it would be *mercury-equals-power*. For although gold was the catalyst and economic pump-primer for California's explosive development, the control of the production, trade, and use of quicksilver was the tool that a handful of elites used to secure their wealth and position as heads of vertical monopolies and to shape a pattern of development serving a wealthy elite. If gold, at least the gold of the California Dream, represents the rugged individualist succeeding by his or her own energy and industry, then quicksilver represents the elite and the self-reproducing power of global capital and concentrated wealth.

Unlike gold, which was valuable because it was money, mercury was valuable for its ability to form an amalgam for gold and silver recovery; without mercury gold and silver were much more difficult to extract. This property made the control of the production, trade, and use of mercury a powerful tool for the British merchant capitalists, their partners, and their successors who established the mercury industry in California. Mercury mining in California predates the discovery of gold by a few years. The first mercury mine, and the first commercial mine of any sort in the state, New Almaden, was just south of San Jose and by 1851 accounted for half of annual global production, disrupting the Rothschild cartel's control of the world quicksilver markets. That California was producing, by 1851, half the world's supply of a commodity as valuable as mercury is evidence of the great power and wealth to be gained through its control. However, unlike the everyman's spectacle of the gold rush, mercury production and trade in early California were controlled by a few very rich and powerful figures who conducted their business of mercury production and trade largely behind the scenes.

The Forgotten History of Mercury Mining

The story of quicksilver mining in California is relatively unknown. Even the knowledge that quicksilver was a major industry in nineteenth-century California—let alone the number-two industry in the value of production through 1890 behind gold mining—escapes most historians, as do the facts that mercury mining was the first mining in California and that the New Almaden Mine was the richest single mine in the history of the state.[2] One of the greatest questions surrounding mercury mining is why the industry is largely forgotten.

Today, mercury is most often mentioned in relation to mercury pollution, a serious and important environmental issue. Few wax nostalgic over the good old days of mercury mining, or celebrate the quicksilver rush of the 1870s. The mythology of the California gold rush, and the phenomenal series of gold and silver booms that followed, dominates the history of mining in California and the West. Although gold and silver are understandable and knowable, most people don't know where mercury comes from or how it is used. Mercury is mysterious and unknowable, and its value comes only from its use. One northern California resort, located in an old mercury mining district, proudly recalls in their promotional literature the "silver" mining that took place there, masking that it was really quicksilver mining that first led to the development of the area, and it was quicksilver mining that created the need for the Superfund site just down the road from the resort.

But despite the many reasons to want to forget quicksilver, not least among these the environmental legacy, there are many reasons to remember it, for it was important in shaping early California following statehood. Mercury was not only the first metal mined on an industrial scale in the West; it was crucial to the industrial-scale refining of the gold and silver mined in the West after the initial placer mining days. That both mercury and bullion were abundant in California is an accident of geology of enormous benefit to California and the American West. Alone, mercury would have been a valuable trade item and given those who controlled its supply some measure of power to control gold and silver supplies elsewhere in the world. In combination with the gold and silver mines of California and the West, the state's mercury was even more valuable and a tool to great power. However, without the presence of mercury, the story of the development of California and the West would have been quite different. If quicksilver had not been mined in California and instead only available from the European mines of Almadén and Idrija, the gold and silver mines of the American West would have been subject, to a significant degree, to the dominance of the European powers who controlled the global quicksilver market. Large quantities of mercury were, however, present in California. The story of mercury in the second half of the nineteenth century, therefore, is the story of prominent and powerful people in the West who fought over the control of mercury, and through it became significant players in gold and silver production, and the development of the West.

Despite being largely forgotten today, quicksilver was recognized as extremely important in gold rush California. The tremendous wealth of the New Almaden Mine was not lost on contemporaries. This 1848 map (Figure 0.1a and detail 0.1b), published in the year of the discovery of gold in California, gives significant weight to the presence of quicksilver in California, and by association marks the importance of one metal to the other. The hills for many miles around what became New Almaden, south of San Jose, are labeled "Quicksilver district," with the description reading "the

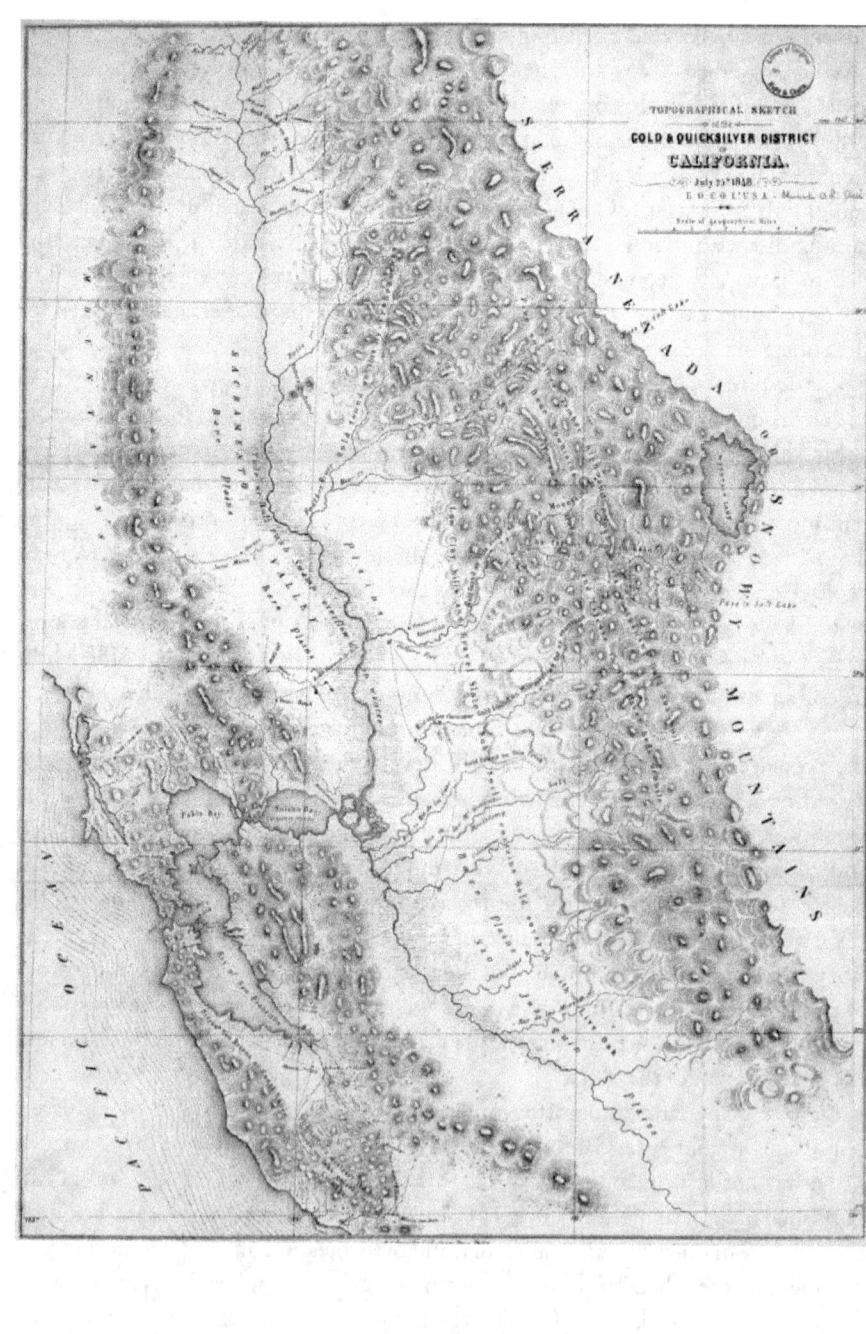

FIGURE 0.1a (facing page) Topographical Sketch of the Gold and Quicksilver District of California, July 25, 1848

This map of the gold and quicksilver district of California was printed in Philadelphia, and for the period viewer shouts of the riches of California. It is a finely etched map that on close reading points out where gold is to be found, mostly in the foothills of the Sierra Mountains. The bottom left corner of the map, below San Francisco Bay, shows the Quicksilver District. "Forbes Quicksilver Mine," later called New Almaden, is labeled with another quicksilver prospect and a silver mine. (Courtesy of the Library of Congress, Ord, *Topographical Sketch of the Gold and Quicksilver District of California*.)

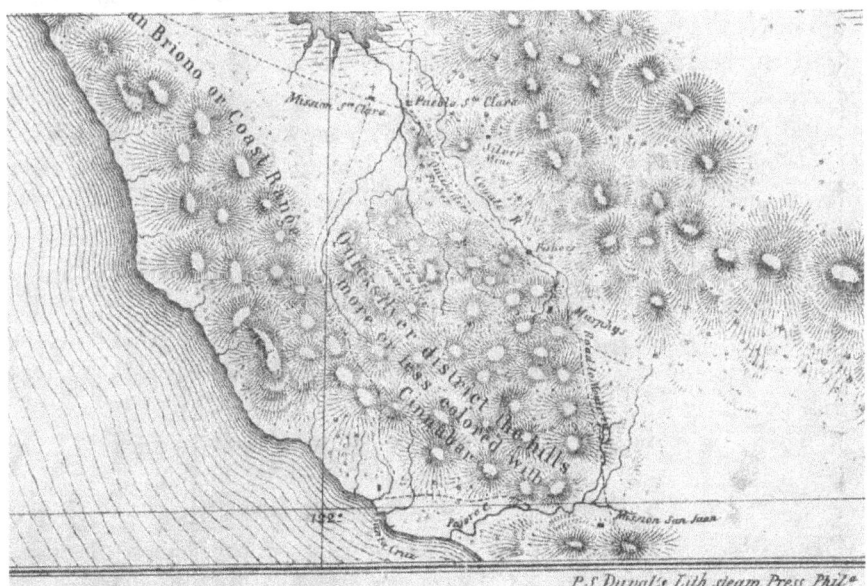

FIGURE 0.1b Topographical Sketch of the Gold and Quicksilver District of California, 1848 (Detail)

This detail shows the Quicksilver District of California centered on Forbes Quicksilver Mine, roughly twenty miles south of San Francisco Bay. The bay is visible at top, just left of center. The hills between Mission San Jose and Mission San Juan are labeled "the hills more or less colored with Cinnabar." (Courtesy of the Library of Congress, Ord, *Topographical Sketch of the Gold and Quicksilver District of California*.)

hills more or less colored with Cinnabar."[3] For the California immigrants this map was meant to attract, a supply of quicksilver bode well for successful gold mining, and was an example of other riches to be had in California.

Mercury mining was an industry of greater importance in the history of California and the West than even its formidable production totals tell—28,000 flasks of about 50,000 flasks produced worldwide in 1851, the first year of reliable production statistics.[4] For the next four decades the California mercury mines produced roughly half of the global supply. The mercury produced in California had a high trade value, and it was successfully sold throughout the Pacific Basin and the world. In fact, half of all California mercury produced in the nineteenth century was exported.[5] But although the sale value of mercury could be lucrative, this value paled in comparison with the wealth and power to be had by controlling the mercury supply. This intimate connection of quicksilver and bullion ended, however, in about 1890, when mercury was supplanted by cyanide as the key to refining gold and silver. In a few short years the western bullion mining landscapes were transformed, as every refining mill was converted from a process of mercury amalgamation to a process of cyanidation.

Gold, silver, and mercury mining were contemporary industries. We know of the mother lode and the other great silver and gold mining districts, but few people know of New Almaden or the other great quicksilver mines of New Idria and Oat Hill, the major quicksilver mining districts throughout the Coast Ranges, or the quicksilver boom of the mid-1870s. The histories of gold, silver, and mercury are intimately tied together, but the ties are not told in the histories of gold and silver—mercury mining has fallen prey to selective memory. The story of quicksilver mining in California and the West has been lost in the all-encompassing fog of the western romance with gold and silver mining.

Gold Mining versus Quicksilver Mining

"The traveller in San Francisco, asking the question Englishmen invariably ask, What's to be seen? would be thus answered. The Big Trees, Eusamity Valley, Napa, and the Quicksilver Mines."

—Charles Dickens, *All the Year Round*, 424–28

In the late 1850s and 1860s the New Almaden Mine was one of the most popular excursions in California for people who could command letters of introduction to the mine. While the gold diggings in the foothills of the Sierra dominated images of mining in popular imagination, New Almaden was the mine people visited, being only a comparatively modest excursion from San Francisco. But more than its physical proximity to the city, the New Almaden Mine offered visitors the experience of the mine of ancient myth, popularly expounded upon by Victorian authors as a labyrinthine world of tunnels and shafts in an underground Hades, populated by mysterious and frightening troglodytes digging fabulously rich ore (Figure 0.2).

FIGURE 0.2 Visitors to the New Almaden Mine, early 1860s
This etching illustrates a story of mine engineer Sherman Day escorting young women (one being his daughter) and their teacher through the mine. Note the pillar of ore left as a support, a mining technique referred to as the "Mexican Method." Note also the large volume of space, presumably the form left after removing the ore. These spaces were called *laborés* at New Almaden. (The etching is by Harry Fenn, a nineteenth-century landscape illustrator and watercolorist. Lucy St. John, "The Quicksilver Mine of New Almaden," 590.)

Probably the most vivid of the traveler's tales is also one of the earliest, that of Mrs. S. A. Downer in 1854. Mrs. Downer wrote in *The Pioneer; or, California Monthly Magazine*, of her trip to New Almaden and descent into the mine. Describing a group of workers she encountered as part of her journey underground, she wrote:

> Another turn brings us upon some men at work. One stands upon a single plank placed high above us in an arch, and he is drilling into the rock above him for the purpose

of placing a charge of powder. It appears very dangerous, yet we are told that no lives have ever been lost, and no more serious accidents have occurred than the bruising of a hand or limb, from carelessness in blasting. How he can maintain his equilibrium is a mystery to us, while with every thrust of the drill his strong chest heaves, and he gives utterance to a sound, something between a grunt and a groan, which is supposed by them to facilitate their labor. Some six or eight men working in one spot, each keeping up this agonizing sound, awakens a keen sympathy. Were it only a cheerful sing-song, one could stand it; but in that dismal place, their wizard-like forms and appearance, relieved but by the light of a single tallow candle stuck in the side of the rock, just sufficing to make "darkness visible," is like opening to us the shades of Tartarus; and the throes elicited from over-wrought human bone and muscle, sound like the anguish wrung from infernal spirits, who hope for no escape.[6]

Mrs. Downer invoked for her educated readers the Greek myth of Tartarus, comparing the New Almaden Mine to the lowest regions of the world; in myth Tartarus is as far below Earth as Earth is below heaven. Tartarus is both a prison for defeated gods (the Titans were condemned here after their defeat by the Olympians) and a place of punishment for sinners. As the visitor perceiving the torment of the miners, Downer placed herself in the role of the hero Aeneas, who in visiting the underworld saw the torments inflicted on those imprisoned there.

As compared with Downer's experience of New Almaden, gold mining in the Sierras, with a preponderance of gravel washing and hydraulicking, offered little satisfaction in fulfilling this ancient myth of the mine and the underworld. The Forty-Niner of popular myth, at the time of the rush, was a white man with little or no training in mining, panning for gold by the side of a stream. The mercury miner, however, was a Mexican or Chilean skilled in the trade who labored hundreds of feet underground in a bewildering maze of tunnels. While the Forty-Niner planned to make a fortune in California and quickly go home, the mercury miner was destined for a short, hard life lived in the mines. The Forty-Niner had a life and a family elsewhere, whereas the Mexican and Chilean miners at New Almaden were part of a community of families who had mined for generations. For travelers of the 1850s and 1860s searching for the mine of myth, the trip to New Almaden often exceeded their expectations, fulfilling their desire to step, for a moment, into a thrilling world unlike their own. But, in addition to the myths writers such as Downer dramatized, the New Almaden visitor encountered physical truths, including the fact that New Almaden was the most developed mine in California, with the most extensive infrastructure and the largest workforce of any single mine in the state. By the early 1850s New Almaden was a fully developed industrial center of global importance, a British/Mexican-controlled corporate island in the midst of the California gold rush frontier.

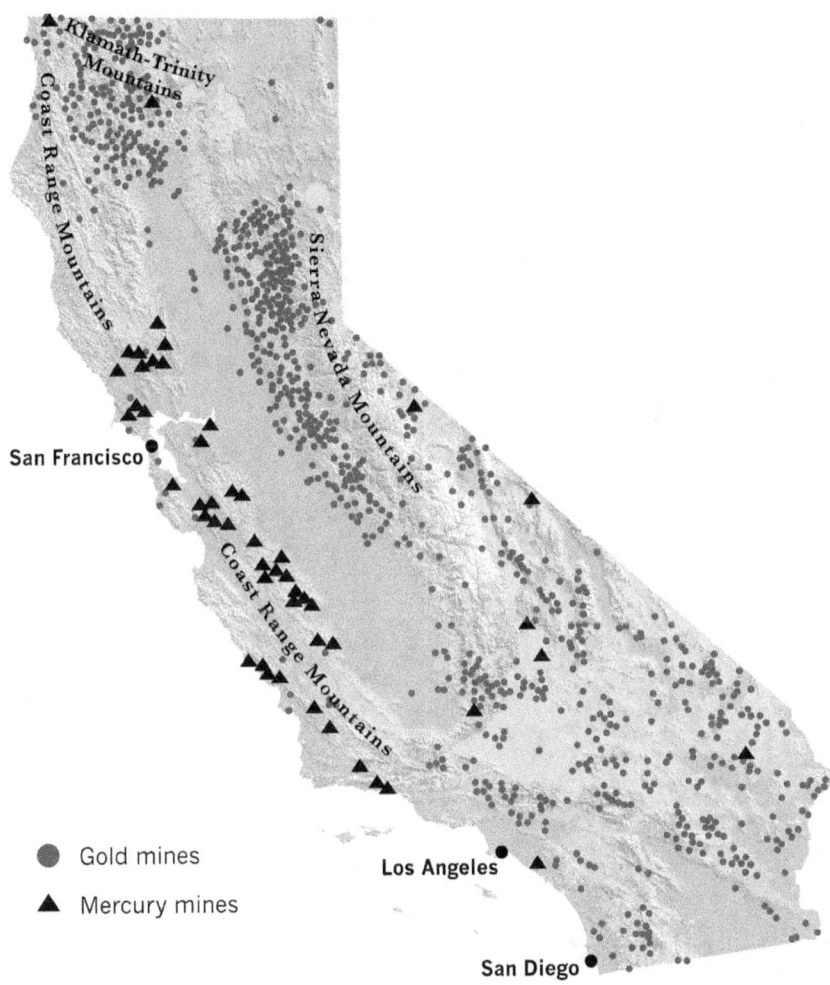

FIGURE 0.3 The Gold Mines and Mercury Mines of California
The gold mines of California were located primarily in the Sierra Nevada Mountains, with later gold mining in the Klamath-Trinity Mountains and throughout Southern California. The mercury mines of California were located in the Coast Ranges, from the Oregon border south to Santa Barbara. (United States Geological Survey; map drawn by Austin Porter.)

The contrast between California's mercury and gold mines, as evidenced by the travelers' tales of the mines, is based in geography and geology. Mercury mining was almost exclusively in the Coast Ranges, from Santa Barbara north to the Oregon

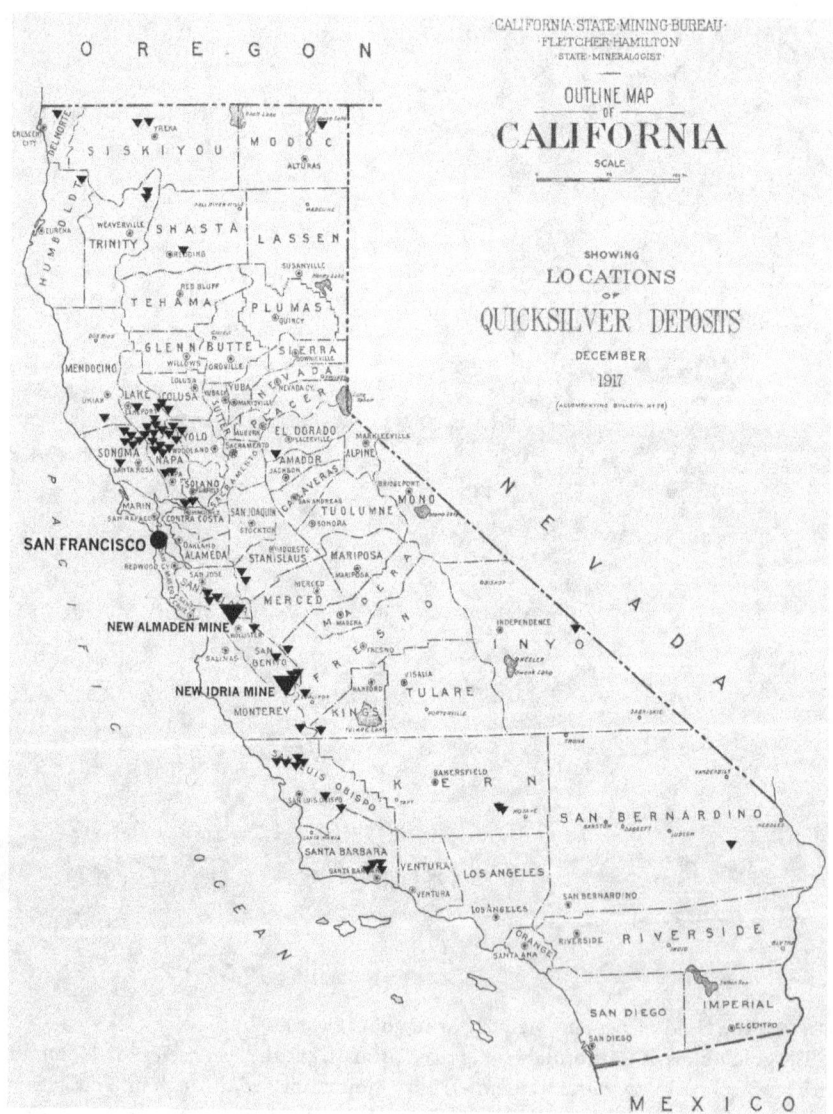

FIGURE 0.4 The Major Mercury Mines of California, 1917

The mercury mines of California were located in the Coast Ranges, from the Oregon border south to Santa Barbara, and are shown on the map as black triangles. The New Almaden and New Idria Mines were the earliest quicksilver mines in the state, and the two largest mercury producers. In the early 1870s rich districts of mercury were developed in Napa, Lake, Sonoma, and San Luis Obispo Counties. (Bradley, *Quicksilver Resources of California Bulletin No. 78*, 108. Map enhanced by Austin Porter.)

border (Figure 0.3). The two largest mercury mines, New Almaden and New Idria, were south of San Francisco Bay; the greatest concentration of mines was in Sonoma, Napa, and Lake Counties (Figure 0.4). Although New Almaden was the richest mercury mine and operated for decades, the second-most productive mercury mine—and the longest-operating mine in the history of the state—was New Idria, which opened in the mid-1850s and closed in 1971. In addition to these two major mines there were dozens of minor mines and 100 or more small mines and prospects that were discovered and developed.

In contrast, gold mining was in the mother lode, an area of 200 by 40 miles in the Sierra foothills and later in other regions such as the Klamath-Trinity Mountains. Gold was found in alluvial deposits in both active and former water channels throughout the whole area. Gold rush miners searched the 8,000 square miles of the mother lode and worked the sand and gravel deposits that looked promising. Alluvial deposits of gold originated in the erosion of gold-bearing quartz veins. Over long periods of time, water (and a variety of other geologic and environmental forces) released gold from these veins and in flowing downhill deposited and concentrated the gold in placer deposits of great richness along the entire length of the belt. Unlike gold, mercury is rarely found in economically viable quantities in its native form. Some panning for mercury is recorded in the Pine Flat Quicksilver District in Sonoma County, but nearly all of the world's mercury has been reduced from cinnabar (HgS), its primary ore. Cinnabar is deposited in veins in rock fissures by rising waters in volcanically active regions.

The primary difference between gold mining and mercury mining in their first decade was that gold mining was placer mining, whereas mercury mining was hardrock mining. In placer mining, alluvial deposits are mined for minerals, and while this form of mining can involve elaborate mining techniques, including tunneling and sophisticated control of water, for the most part placer mining involved the classic Forty-Niner locating a promising sandbank and panning for gold, or using a rocker box and a shovel to recover bits of pure metal. Hydraulic mining is a form of placer mining practiced in California from 1853 to 1884 that introduced new efficiencies into the industrial recovery of alluvial gold by blasting large areas with powerful streams of water, though at great environmental cost.[7] Cinnabar mining, in contrast, is a form of hardrock mining in which hard minerals are removed from underground deposits.[8] Hardrock mining involves elaborate underground infrastructure, including tunnels, shafts, lifting equipment, and skilled mining techniques such as blasting and supporting underground workings. As opposed to placer mining, which can be accomplished very cheaply in its basic form, the elaborate underground workings of hardrock mining demand significant capital outlay and a skilled, specialized workforce.

FIGURE 0.5 Quicksilver Machine, in Mormon Gulch

This machine used mercury, water, and a rocking motion to separate gold from other soil components. Miners added mercury to the machine, and due to its heaviness it sat between cleats along the bottom of the water path through the machine. Gold-bearing soil was then washed through the machine and agitated with a rocking motion. Being heavy, the gold particles sank to the bottom and there amalgamated with the mercury. The mercury and gold were then separated in a furnace. (William Wells, "How We Get Gold in California," 605.)

This difference between gold and mercury mining, of placer versus hardrock mining, was fundamental to the very different forms of the gold and mercury mining industries from the late 1840s to the late 1850s and the subsequent development of California. Whereas gold was a poor man's metal, in theory at least accessible to any able-bodied person, quicksilver was a rich man's metal, requiring large capital outlays for development and an industrial organization of labor.[9] The placer deposits of alluvial gold meant that anyone with simple tools had a chance of finding gold and making money. And because the gold district was so large, there were plenty of places for the tens of thousands of gold seekers to try their luck. The only capital required for placer gold mining was on the scale of the individual miner: transportation to the goldfields and money for tools and supplies. The individual miner was an economically viable unit of production. In fact, many miners joined together in small companies of men, but again the amount of capital required for these miners to be productive was quite small. It is notable that the skills required for placer gold mining were basic and could be learned quickly and practiced without much bodily risk (Figure 0.5).

While the individual miner and his small amount of capital, repeated tens of thousands of times, was the basic economic unit of the gold rush, one very powerful and well-capitalized British merchant house, Barron, Forbes & Co., dominated the mercury industry. Barron, Forbes & Co. first leased, then bought, the New Almaden Mine along with a few partners, including other merchant houses and well-placed powerful figures in early California.[10] This group was called the New Almaden Company, with Barron, Forbes & Co. as the majority partner. The rich cinnabar deposits of New Almaden were concentrated on one hill just south of San Francisco Bay. Extracting the cinnabar required hardrock mining, and the New Almaden Company brought skilled miners from Mexico and Chile to work as wage labor (or perhaps in bondage or peonage) in the development of the mine. So, whereas gold mining in California began as a mad rush of tens of thousands of individuals, all working for themselves or in small groups and using simple hand tools, quicksilver mining in California was controlled for the first twenty years by a single major British trading house employing wage labor and using industrialized mining techniques, exploiting a fabulously rich mine with ore averaging over 30 percent mercury during the decade of the 1850s.[11]

Quicksilver was harder to mine and process than was gold, which required no processing. Quicksilver required capital to construct a mine and a processing plant in order to produce a marketable product. Mining at New Almaden, however, was relatively easy at first, with rich ore near the surface. Furthermore, cinnabar was the easiest metal ore to process, requiring only heating and cooling to vaporize and then condense the mercury. New Almaden dominated the California mercury industry, and until the time of the American Civil War the industry in California was almost entirely synonymous with the largely non-American-owned New Almaden Company.

The different geology of gold and mercury led to each having different structures of the use and ownership of land. Placer mining claims were relatively simple: a claim was defined by a number of running feet—often along a stream or waterway—and once a claim was made, the claimant worked the soils within its boundaries to extract the gold. Tens of thousands of these small claims were made by miners in the goldfields.

The rule adopted in California's gold camps was "one miner, one claim," and rights to that land were not based on ownership, but on discovery and use. In fact, these small-time miners resisted attempts to convey ownership on mining claims. Beyond this basic right to claim mineral wealth on public lands in California, the miners in each mining district formed a miners' convention to establish localized rules for laying claim to a mine (where the claim had to be filed, size of claim, etc.) and for holding that claim. A set of basic miners' conventions came to be commonly accepted throughout the state.[12]

Quicksilver, however, was found in veins and ore bodies embedded in rock underground. Claims for this type of ore were much more complicated to define by boundaries, since U.S. mining law held that the original claimant of an ore outcropping could follow the vein wherever it led, a potential nightmare given the complicated geologic structure of cinnabar. Some quicksilver mining districts were organized as "claims" on a gold mining model, such as the Pine Flat District. These claims were made during the quicksilver rushes and few if any ever amounted to much. Most claims, however, were based on ownership, and in many cases this ownership was based on Mexican land grants made when California was part of Mexico. Reconciling these claims with American law was highly contentious. For example, in the late 1840s the New Almaden Company established itself as the only quicksilver power in the state, with a Mexican land grant encompassing a large tract around the mine. The people involved with the New Almaden Company successfully suppressed quicksilver prospecting in the state by buying other prospects. If a competitor did not sell out and actually produced marketable quicksilver, they may have had a difficult time selling it due to Barron, Forbes & Co.'s control of mercury markets. In this situation would-be mercury prospectors had almost no chance at success.

The great differences between gold mining and quicksilver mining extended, not surprisingly, to the people involved in the industries. Gold miners were largely adventurers from around the world learning the simplest of mining methods on the job and with no history of solidarity with other miners. They rushed to California to get rich and many returned home. In contrast, the workers at New Almaden—at first Mexicans and Chileans—were experienced miners from silver mines in Mexico and Chile, with developed mining skills and a long history as a mining class. Unlike gold mining, quicksilver mining in California had a labor force with a multigenerational history of mining under the Spanish Empire.

In 1850 the Foreign Miners' Tax, which charged "foreign" miners in California twenty dollars a month for the right to mine, made placer mining an increasingly difficult livelihood for Hispanics and nonwhites. Recommended by the California Senate Finance Committee, this tax was enforced largely along racial and ethnic lines, with Chinese, Mexicans, and Central and South Americans bearing the brunt of the tax. This tax aided the New Almaden Company in keeping their Mexican and Chilean workforce. New Almaden became an island of acceptance for these men and their families in a state that was growing increasingly hostile to them. Unlike the single male gold seekers, the miners recruited to New Almaden came with their families with plans to make their lives there. Undoubtedly the New Almaden Company benefited from the increasing institutionalization of racism in the state, which provided them a stable and skilled workforce that had few options for improved conditions elsewhere in the state.

This stable workforce of the New Almaden Company lived and worked in the New Almaden hacienda, the Spanish Empire plantation-like system at the mine. The mine was developed on it, and it, in turn, organized the mine. This is in contrast to the gold rush settlements that ran the gamut from tent "cities" to speculative towns. Mining camps in the goldfields were as permanent or as temporary as the supply of ore. Most camps were composed of tents that people set up near their claims. Such camps lent themselves to the average gold rush miner's state of "permanent impermanence." Within the mother lode, there were also prospective towns laid out by optimistic speculators. People could buy lots in these towns and build on them as they saw fit. A handful of these towns flourished, with temporary structures giving way to more permanent structures as economics improved. Most of these towns, however, failed as miners moved on to the next gold strike.

In contrast, the quicksilver camps in early California were built as permanent or at least semipermanent planned communities on the hacienda model. At New Almaden from the earliest days there were two distinct settlements: a formal village built near the reduction works in a valley below the mine, and "Spanishtown," the miners' camp, built near the summit of Mine Hill near the main entrance to the mine. These settlements were both on company property, and wages from the mine supported the whole community. Spanishtown was mostly built by the miners in piecemeal fashion, and consisted of houses and a few stores, saloons, and other services organized by a wide street functioning as a plaza. The formal town, however, was a showpiece for the mine, featuring by the mid-1850s one of the largest manor houses in the state, the Casa Grande, and a picturesque row of managers' cottages along a tree-lined street with a diverted stream running along it. At the end of this alley were company offices, a store, a hotel, and the reduction plant, which for most of the 1850s was the most advanced mineral-processing facility in the state. When miners in the goldfields were living in tents and speculative boomtowns, the miners at New Almaden were living in an expansive industrialized settlement that stood out in early California because of its size, permanence, and control by one company.

Quicksilver and California

This book explores how focusing on mercury changes our understanding of the development of California and the West by examining the mercury industry and its landscapes in detail. *Mercury and the Making of California* explores the mercury mining industry and its role in the development of the state. The thread of the mercury mining industry in the tapestry of California history provides us a theme for study, and a transect for research and understanding across five decades, from the 1840s to the 1890s. Mercury itself—and its production, trade, and use—comprises a "mercury system" with intermeshed physical and social components. By looking closely at the resource (mercury), we can see power relations and how they functioned.

The first part of the book—chapters 1, 2, and 3—focuses on the mercury industry in California, including its geography, how it was embedded in world systems, and its centrality to systems of power and wealth. These chapters explore the idea of transformation—how a Spanish and British imperial system of mercury production, trade, and use that supported state powers from the sixteenth to the nineteenth century was transformed in rapidly Americanizing California into a racialized, American capitalist system of the production, trade, and use of mercury, leading to great wealth and power for a few. The second part of the book—chapters 4, 5, and 6—looks at the changing landscapes of the mines and camps of the mercury industry, focusing on issues of work and family life, and the importance of race and racial hierarchy in the social and physical construction of these landscapes. Together, the six chapters tell a multidimensional story of California quicksilver.

The story of mercury mining in the American West begins with a story of British imperial expansion, in the guise of British merchant houses, expanding up the South American and Central American Pacific Coast into California. These merchant houses, building on growing global British economic dominance while feeding on the decaying remnants of the Spanish Empire, brought with them systems for extracting wealth from the production, trade, and use of California's mercury, establishing the industry in California while it was still part of Mexico. Chapter 1 argues that mercury mining in California was a hybrid of Spanish and British colonial structures of mercury production, trade, and use on the California frontier. With the opening of the New Almaden Mine in the 1840s, California quicksilver became the latest chapter in a long, global drama of interrelations between mercury and humankind. California, however, was undergoing rapid transition, and the state's burgeoning mercury industry—developed in the legacy of the Spanish and British Empires—met with the rapid Americanization of the state and underwent transformation. How the new Californians creating the mercury industry in the state combine the Spanish and British imperial legacy with the rapid Americanization of the state to use mercury as a tool for acquiring phenomenal wealth is an important and untold story in the development of the American West, a very different story from the story of gold and silver mining.

During the American Civil War the ownership of the New Almaden Mine was wrested from the British merchant house by American interests, largely from the East Coast. Despite losing this fabulously rich mine, the business descendents of the merchant house—themselves now Californians—used their connections and knowledge of the industry to form a range of partnerships to control the production, trade, and use of California's mercury, and to become immensely rich in the process. Chapter 2 tells this story of the money and power in the industry, arguing that the mercury mining industry—begun as an industrial tool of empire—was transformed into a tool of state building by powerful Californians scheming for the control of mercury resources and the wealth possible through this control.

Although mercury was a powerful tool manipulated by elites to gain wealth and power, it also figured large in the lives of many others in the state, particularly those who worked in the industry, and their families. Understanding the impact of quicksilver in the development of California involves exploring the everyday world of the mines and camps of the mercury industry, for it was in these landscapes that struggles between groups of people took place. Interpreting the everyday landscapes of the industry involves looking in a combination of ways at both as mercury as a resource, and how groups of people struggled to exploit the resource.

As a valuable resource, quicksilver was fought over and contested by various groups of people for their own purposes. Capitalists of various guises; workers of various races and ethnicities; and traders, middlemen, and many others tied into the social webs of quicksilver production, trade, and use employed quicksilver to make the world to their liking. Change in the production, trade, and use of California quicksilver occurred as these groups were more or less successful in using quicksilver to achieve their goals. This book tells the story of these various groups, their interrelations and power struggles, and the meanings of these contests for the development of California.

To this end, chapter 3 presents an integrated geology and a geography of mercury mining in California. Different groups of people involved with the mercury industry—including capitalists and workers—understood that cinnabar could be used to create wealth and power, and each group struggled to secure a piece of the wealth. Mining was the means by which these groups exploited the cinnabar resource to create wealth, and thus the landscapes of the mines are central to this story. The various groups of people involved in the industry imagined the world differently from each other, and the mines—including the landscapes of work and the landscapes of everyday life—

were the physical places where these groups acted out their different worldviews and their struggles. Chapter 3 concludes by presenting a typology for understanding the spaces of the industry. These types are each a culturally negotiated tool for exploiting mercury. Interestingly, during the quicksilver boom of the 1870s, there were four distinct types, or models, for running a mercury mine successfully operating in the state. Each of these types had a powerful racial and ethnic component that in many ways dictated the physical form of the mines and camps, and the way that mining work was done at the mines. This racial component in the everyday landscapes of the mercury industry is essential to understanding quicksilver mining and how it differed from other mining industries in the state. Interestingly for the history of the development of the state, the structure of race in the mercury industry was most akin to structures of race in the state's agricultural and factory production that took form decades later.

As a linked set, the first three chapters of this book explore the industry as embedded in a world system and consider the creation of wealth and power in the state. This first section of the book also introduces the everyday landscapes of the mines and camps, exploring the day-to-day experiences of the thousands of people working in the industry. I argue that understanding questions of race and racial hierarchy are crucial to understanding the quicksilver mining industry and its role in the development of California. Within the industry, race influenced factors such as work and the application of technology in mining and reducing mercury. Moreover, race and ethnicity influenced family life and camp life at the mines. Geographically, across the sites of the California mercury industry, race and ethnicity influenced patterns of development.

Chapters 4, 5, and 6 explore in detail the landscapes of the mercury mines of California, how they were adapted and changed over time, and what the form of the mines and how they changed say about the development of California. These chapters are roughly chronological. Chapter 4 explores how both the underground and aboveground landscapes of work and the landscapes of camp life at New Almaden were shaped by the ethnic and racial hierarchies at the mine, focusing on the period of transition in the 1860s, when the mine was taken over by American owners. Chapters 5 and 6 are a detailed exploration of the mines and camps of the industry during the quicksilver boom of the early 1870s. As mentioned above, during the boom there were four coexisting types or models of mercury mines, and although each type was vastly different in the racial structure of the industry workers and in the physical form of the mines and camps, all four maintained the same racial and ethnic hierarchy prevalent throughout the state. Chapter 5 details how work and the application of technology were shaped by race and ethnicity, and chapter 6 considers the physical landscapes of the company towns and camps and how they were shaped by race and ethnicity.

The conclusion of the book makes the case that California is the Quicksilver State. Within the particular history of mercury mining in the state are embedded truths about the development of the state that show us a very different California and tell a different story of the development of California than does gold or silver mining. The story of California quicksilver shows California to be not only a place with a wealth of natural resources for exploitation, but also an industrial place with ties to global finances and trade from pre–gold rush days. California quicksilver tells a story of European imperial expansion as well as American nation building, and together these ideas create a uniquely Californian experience. California is the Golden State, and always will be. However, California is also the Quicksilver State.

Notes

1. Carey McWilliams, *California: The Great Exception*, 25.

2. For quicksilver as the number-two industry in value of production, see Andrew Isenberg, *Mining California*, 165. For New Almaden as the richest mine, see Robin W. Winks, *Frederick Billings*, 99.

3. Cinnabar is the primary ore of mercury.

4. Flasks contain seventy-six pounds of mercury, a measure that comes to us from Roman times. For production totals see Walter W. Bradley, *Quicksilver Resources of California Bulletin No. 78*.

5. David J. St. Clair, "New Almaden and California Quicksilver in the Pacific Rim Economy," 278–95.

6. S. A. Downer, "On Her Trip into the New Almaden Mine," 222–23.

7. Hydraulic mining in the state was greatly curtailed by the *Woodruff v. North Bloomfield Gravel Mining Company* court ruling in 1884.

8. Later, following the placer period, gold was mined in various ways, including hardrock mining. Gold typically occurs in quartz veins, and the extraction of gold from its original veins is called quartz mining.

9. See McWilliams for a full discussion of gold as a poor man's metal. McWilliams, *California*, chap. 3.

10. John Mayo, *Commerce and Contraband on Mexico's West Coast in the Era of Barron, Forbes & Co., 1821–1859*, 390–94.

11. By comparison, the New Idria Mine was operated for much of the twentieth century with ore yielding less than 4 percent mercury.

12. Charles Howard Shinn, *Mining Camps*.

1
Imperialism and California's Quicksilver

Years before gold was discovered at Sutter's Mill, quicksilver was being produced at the New Almaden Mine in the hills a few miles to the south of Mission San Jose. The first mine in what was to become the state of California, New Almaden was recognized at the time for its potential value, and the mine was much discussed. Thomas Larkin, the U.S. consul at Monterey, wrote to Secretary of State James Buchanan regarding New Almaden on March 28, 1848, not long after American annexation of California and not long after accounts of the discovery of gold in the Sierra foothills appeared in San Francisco newspapers:[1]

> Messrs. Barron Forbes & Co. of San Blas and Tepic, one of the richest English Houses in Mexico, in 1846 became lessees for sixteen years of a quicksilver mine... Since that period they have become part owners. They have had a few common labourers, with some pick axes, crow bars, shovels and common whaling try-pots... everything done very imperfectly. Nevertheless, they are now taking out two hundred pounds per day... Some of the ore now found is one half pure quicksilver... Several mines of quicksilver have in 1847–48 been discovered in this territory. From every appearance California will soon supply all of Mexico and South America and be able to undersell any mines in the world.[2]

Larkin was a successful businessman in California, and had been firmly in favor of American annexation. While he was the U.S. consul, tasked with representing the United States in California, he was also made, in 1845, the "confidential agent in California" in a secret dispatch from Buchanan.[3] As confidential agent his job was to promote U.S. interests in California against the designs of other governments, particularly the British, and this he had done for years, reporting to Washington any rumor of British involvement in the region. Larkin was aware of the value of quicksilver mines, and the New Almaden Mine and other quicksilver mine prospects in California occur many times in his letters.[4] While Larkin was acting in the interests of

DOI: 10.5876/9781607322436:c01

FIGURE 1.1 The Reduction Works at New Almaden, drawn between 1847 and 1851

The reduction works at New Almaden as drawn by Frenchman Fritz Wilkersheim, who toured in California sometime between 1847 and 1851. The caption, written in French with the Spanish word *asiento* (meaning establishment or settlement) in parentheses, reads, "California—establishment of the mine of mercury of New Almaden." This is a very early image of the mine works at New Almaden. Furnaces are seen in the trees at the far left of the image. Another furnace is at the rear of the yard with a condensing system working up the hill. In the foreground is the bowed-roof building prominent in many early images of the works. On the hillside many trees are showing the effects of the mercury. (Courtesy of the Robert B. Honeyman Jr. Collection, 1963, the Bancroft Library, 002:1304:31-ALB.)

the United States government, he continued to act as a businessman for his own benefit, attempting through numerous channels to acquire shares in the New Almaden Mine and other quicksilver prospects in California.[5] About the time of his letter, Larkin formed a mining company with seven partners, one of whom held a claim for mining rights in the New Almaden area.[6]

In the letter quoted above, Larkin's goal was to convince the powers in Washington that the recent annexation was an opportunity to legally unseat the powerful British interests controlling the New Almaden Mine. In the letter he describes Mexican law regarding mines and the changes being made to laws regarding mining in California by the recently appointed military governor, the combination of which he saw as an opportunity for invalidating the Barron, Forbes & Co. title in favor of U.S. interests.[7]

FIGURE 1.2 The Usine (Mill) at New Almaden, drawn between 1849 and 1852

The reduction works at New Almaden as drawn by Paul Emmert (Swiss, 1826–67), who worked in California between 1849 and 1853. Reduction of ore at this time involved intermittently fired furnaces such as those pictured with smokestacks to the left of the image. Photographs from the 1850s and 1860s, especially those by Carleton Watkins from 1863, show similar furnaces and stacks (see Figure 1.9). (Courtesy of the Bancroft Library, 1974.002:BFR.)

His claim that at New Almaden "everything (was) done very imperfectly" is curious, given that within eighteen months, the mine was producing enough mercury to double world production.[8] Also, it is not likely that Barron, Forbes & Co.—the dominant merchant house on the Pacific Coast of the Americas, with experience running other mines and mills—would develop and manage so poorly a rich prospect that they had spent a great deal of money acquiring (Figures 1.1 and 1.2).[9] More likely, Larkin is helping to make his case for taking the mine by claiming mismanagement. And, if in the process of unseating Barron, Forbes & Co. the mine should somehow fall into the hands of his own mining company—well, so much the better, in his view. A decade after Larkin's letter, Buchanan's administration aided in invalidating the title to the mine, an important step in taking the mine from its owners; and eventually the U.S. Supreme Court, aided by the Lincoln administration, wrested ownership of the New Almaden Mine from Barron, Forbes & Co. in 1863, during the height of the Civil War. Taking the mine from Barron, Forbes & Co. was a fifteen-year process, and in the 1840s when he wrote his letter as confidential agent, Larkin stood at the nexus of imperial designs in California, fighting for American interests against those of the British and Mexicans, all the while angling for his own business and financial interests.

Although Barron, Forbes & Co. did eventually lose ownership of New Almaden, the merchant house dominated quicksilver production in California from the late 1840s until 1863, defining the industry and establishing the physical landscapes of mercury mining and reduction in California. Barron, Forbes & Co. understood the potential value of mercury in the Americas, and after leasing the mine quickly worked to develop New Almaden and exploit the quicksilver resource. In 1851 the mine produced 27,779 flasks of mercury, doubling the typical year's production of mercury available on world markets. This scale of production required a landscape of industrial production, and Barron, Forbes & Co.—along with other partners in the New Almaden Company—financed this development, including the mining and reduction facilities and settlements for the workers, managers, and their families. Due to the incredible richness of the cinnabar deposit; the substantial wealth, knowledge, and connections of Barron, Forbes & Co. and its partners; and the work of the laborers, New Almaden developed very rapidly in the late 1840s and early 1850s, establishing itself as the third major quicksilver producer in the world after the Almadén Mine in Spain, operating since Roman times, and the Idria Mine in modern-day Slovenia, operating since 1497. Barron, Forbes & Co. and its partners understood the long historical relationships between mercury and empire, and they developed the mines and settlements of New Almaden to exploit their understanding and create their own business empire from the promise of California's mercury wealth.

Hacienda Nuevo Almaden

The landscape of New Almaden, as established and developed in the 1840s and 1850s, was a hybrid of Spanish, British, and Mexican influence in California. Embedded in the landscape is the story of the development of the industry in California. Located in a region of steep hills and narrow valleys about eighteen miles south of San Francisco Bay, New Almaden was originally a mining grant awarded by the Mexican government to Andrés Castillero, a Mexican citizen who identified and claimed it as a quicksilver mine. The original mine at New Almaden was at the top of what was known as Mine Hill, at an elevation of over 1,700 feet, with commanding views over the bay and the San Juan Valley. The surrounding hills and high valleys around Mine Hill were the sites of additional mine workings and of miners' houses. Figure 1.3 shows the sequence of landscape features the visitor would have encountered in entering the valley of New Almaden. There were a small Protestant church for the owners and managers, the Casa Grande and its extensive gardens and vineyards, and

FIGURE 1.3 New Almaden Hacienda, 1863 (Detail)
Following the road into New Almaden, from lower right, the visitor encountered a Protestant church; the Casa Grande and its extensive, landscaped grounds; the row of managers' and reduction workers' cottages; and in the distance the reduction works. Mine Hill is off the photo to the upper right. (Photograph is by Carleton Watkins. Courtesy of the California History Room, California State Library, 2010-0637; there is a similar but earlier view in Michael Boulland and Arthur Boudreault, *Images of New Almaden*, 42.)

a tidy row of managers' cottages, and at the end of the valley was the reduction works, where half of the world's quicksilver was reduced from the cinnabar ore mined on Mine Hill.

New Almaden, and the fledgling California mercury mining industry as it was created in the 1840s and 1850s, were but a snapshot in the long history of the interplay between humans and mercury. The New Almaden landscape, a hybrid of colonial influences, was at its core a Spanish/Mexican hacienda, or estate, on the California frontier, and it was through the hacienda model that mercury wealth was created and controlled.[10] The hacienda was a colonial economic model imported to the New World by the Spanish in the sixteenth century and a model that remained intact in Mexico until the Mexican Revolution of 1901–17. Haciendas had many similarities to, and important differences from, plantation systems established by other colonial powers in the New World. Haciendas were efficient, industrialized centers that focused on the production of one commodity or product—be it farming, ranching, or mining—as was the case with New Almaden. A hacienda aimed to be an isolated, self-sufficient, interdependent society under the rule of a "patron" and managed by a small circle of elites. The laborers on haciendas worked in a system of peonage, receiving basic food and shelter for them and their families but having little control over their working situation or much of a possibility of mobility. Although haciendas were established for efficient economic production, they were marks of status for their patrons and the other elites, displaying wealth through large landholdings, grand architecture, and luxurious consumption. Haciendas functioned as "inns" for elite travelers as well—hosting and entertaining important guests was an important role for the patron of the hacienda, and many of the travelers to New Almaden remarked on the graciousness of their hosts.

The Hacienda Nuevo Almaden was large and complex, sharing many similarities with the ideal hacienda. New Almaden was built for the profitable mining and refining of quicksilver, but it was also much more than the basic facilities needed to physically mine and process ore. It was consciously created as a separate community, an isolated island in rapidly Americanizing California. Barron, Forbes & Co. had used the hacienda model before at other mines and at the merchant house's cotton mill outside Tepic, Mexico. It was a contained, self-sufficient machine for producing wealth and power while resisting the buffeting of the revolution and change of the era. Like the ideal hacienda, there was a show quality to much of the landscape of New Almaden, highlighted by a grand house and gardens where elite visitors were entertained. Like haciendas built throughout Spanish colonies in the Americas, New Almaden was built on the backs, especially early on, of a workforce that perhaps was held in peonage, limited in opportunity by racial and ethnic bias in early California.

New Almaden differed from the ideal hacienda in important respects. First, it did not produce a raw material or a product for a home country. New Almaden produced a finished commodity that was sold throughout the Pacific world by the Barron,

Forbes & Co. trading network. Unlike the ideal hacienda, there was no single patron of Hacienda Nuevo Almaden, but instead a circle of elites, including the owners and their managers, their lawyers, and businessmen and politicians who were their allies. Eustace Barron and Alexander Forbes, the partners of Barron, Forbes & Co., did not live at New Almaden—in fact, Eustace Barron never visited California, and Alexander Forbes only visited New Almaden in its earliest days. New Almaden was, instead, a grand hacienda for a body of interested elites, who visited the mine and partook of its hospitality as needed and desired. New Almaden was also a hacienda with no direct peers, and in some sense a place without a country. The Hacienda Nuevo Almaden was a hybrid form, adapted to developing California by its owners, managers, and workers to exploit the mercury that promised each group what they needed or desired.

A Visit to New Almaden in 1854

Profoundly influenced by its Spanish and Mexican roots, the Hacienda Nuevo Almaden was a unique colonial implantation in American California. Founded in 1846 in Alta California, it was resolutely its own island of curiosity to outsiders, fundamentally different from gold rush California. It was, perhaps, this foreignness that made New Almaden an attraction for early travelers in California. Stories of the great wealth of the mine also made New Almaden a place of great interest in early California. Early travelers saw the physical landscape of New Almaden and agreed that it was unlike anything else in the state. One of the earliest and most extensive accounts of the New Almaden Mine was written in 1854 by Mrs. S. A. Downer for *The Pioneer; or, California Monthly Magazine*.[11] Entitled "On Her Trip into the New Almaden Mine," it is an article-length account published in the first magazine solely devoted to California, introducing the new state to a readership in California and the East. Downer was a Victorian-era woman travel writer, and presented for the genteel reader of the 1850s not only what you would see at New Almaden, but also how you were supposed to see it.[12] Traveling with Downer to New Almaden as it was developing as a hacienda in California in the 1850s provides a unique view on how a developing power elite in California saw the mine and its significance in California.

Mrs. Downer's story begins on the road from San Jose to the mine. Like most elite visitors, she arrived at New Almaden by stage from San Jose, after a boat ride down San Francisco Bay, or a stage ride from San Francisco to San Jose, we don't know which. She remarked on the beauty of the San Jose Valley and the magnificence of the Coast Range mountains in which the mine lay. This interest in the natural world, from the large-scale geography of the dramatic setting of the mine to the gooseberries and wildflowers growing along the trail, is a dominant theme throughout her article.

FIGURE 1.4 Ranch of the Chief Engineer, 1856
Etching of the Hacienda village, showing the formal aspects of the cottages and tree-lined street. (J. Ross Browne, "Down in the Cinnabar Mines," 545–60.)

For Downer, nature was to be appreciated in its wild state, but it was also to be controlled and put to man's use. On the second page of her article she lamented that until the present company was formed, "man was too timid to avail himself of her gifts," referring to the taking of the great richness of the mine, and all of the timber, water, and other resources at hand for exploiting the abundant ore.

The first buildings she encountered at the mine were a carefully tailored row of cottages for the superintendents of the works (Figures 1.4 and 1.5). These cottages were placed formally on the main road that led from the entrance to New Almaden valley to the reduction works at the other end of the valley. Downer wrote that "each [cottage is] enclosed with a paling fence, containing in front a small flower-garden with shrubbery, while a vegetable garden in the rear bespeaks usefulness combined with taste and beauty."[13] She commented that New Almaden village seemed a moral place

FIGURE 1.5 New Almaden Hacienda, 1863
 This photograph, by Carleton Watkins, shows the neat rows of cottages lining the main road leading to the reduction works. The orderly cottages and rhythmic sycamores convey a strong, formal message, emerging from the Spanish/Mexican hacienda planning established by Barron, Forbes & Co. (Courtesy of the California History Room, California State Library, 2010-0633.)

with well-treated managers, supervisors, skilled tradesmen, and their families living in permanent and picturesque homes. Mrs. Downer stressed this theme of morality for her readers, and the wives and families and beautiful arbors were signs of the morality of the owners and managers of the mine. The village provided the perfect scenery, supportive both of the ideals embodied in New Almaden as a hacienda and of Downer's ideal of morality.

From the village, Downer traveled up to Mine Hill, over a thousand feet in elevation above the valley floor. Downer remarked on the thrill of the precarious journey up the steep mountain roads, and the sublimity of the view and bodily exposure on the hill. Once on the hill, she was impressed by the industry and the scale of the

FIGURE 1.6 The Village du Mineurs at New Almaden, drawn between 1849 and 1852
The miners' village at New Almaden as drawn by Paul Emmert, who worked in California between 1849 and 1853. (Courtesy of the Bancroft Library, 1974.002:6-FR.)

workings, and she remarked on the three major elements of the Mine Hill landscape: the Mexican village, the mine, and the patio or sorting floor for receiving the ore from the mine.

On the hill, Downer encountered many of the hundreds of miners and sorters who both lived and worked there. The Mexican village where they lived with their families, however, was only mentioned in passing and was not visited (Figure 1.6). Downer gets close to but never quite sees the village, describing it as over the next rise and as containing the lodging cabins of the miners and their families and a store for the miners "who may truly be said to live hand to mouth."[14] In Figure 1.7 we can see how, in contrast to the lower village, the Mexican village was in a windswept, exposed location with dramatic scenery but with no water and little shade. For Downer, these workers and their families were peons, laborers with little control over their employment conditions. Visitors did not write of how they lived, for this knowledge would expose some reality of the mines perhaps best left unsaid by the genteel traveler. The worker was of no real account except for the work he performed, which to the visitor was interesting, dangerous, and inhuman despite being performed by humans who with their families lived with this work.

The mine was at the core of Downer's, and every other visitor's, fascination with New Almaden, offering visitors the chance to enter the mythical land of the under-

FIGURE 1.7 Spanishtown, New Almaden Hill, 1863
This photograph, by Carleton Watkins, shows Spanishtown from Mine Hill. The Catholic church and the plaza are just to the left of the center of the image. The main *planilla* (sorting floor) and Englishtown are off to the bottom left. (Courtesy of the California History Room, 2010-0644.)

world, with strange and frightening environments to see and experience. The miners fascinated Downer, but were dealt with abstractly, compared to characters in ancient mythology (as we have seen in her quote in the introduction), and celebrated for both the inhumanly burdensome work that they did and their simple and instinctual release when the work was done. Downer never actually spoke to the miners, or told their stories. She did, however, vividly place them in the underground landscape of the mine, this landscape depicted as a cross between the factories of early industrialization and the underworld of myth. Downer attempted to portray the mine as a factory underground, carved from the earth, yet she could find little logic or system in its construction and operation, feeling instead wonderment, disorientation, and fear.

Following the ore out of the mine, Downer described the patio, or ore-sorting floor, at the mouth of the mine. She presented the patio as a large and elaborately choreographed work area that was the key to the contract mining system used at New Almaden, because it was on the patio that the value of the ore, and thus the pay to workers, were decided. As Downer explained, in the contract system, groups of workers bid on a contract to do an amount of work specified by the company. The ore from each company was kept separate on the patio, where it was graded and weighed with representatives of competing interests all present. Downer detailed the activities of the patio in her article, in part because the patio was where the wealth of the mine became visible (Figure 1.8). More important, though, Downer presented the patio as the intersection of the world of the managers and owners and the world of the miners, asserting that the contract system was fair for all, allowing companies of workers to be paid and judged based on their own merits gauged by production. Downer understood and approved of the patio—it was where the underworld and its minions gave way, in her mind, to the forces of light and morality.

Downer followed the ore from the patio to the mine reduction works down in the valley, a thousand feet below (Figure 1.9). It was at the reduction works that the cinnabar ore was reduced, that is, heated in furnaces until the mercury vaporized. The resulting mercuric gas was then cooled in large condensers where it condensed and was collected. Interestingly, the reduction works were referred to as the "hacienda" for much of the mine's history, in this case the term coming from "Hacienda de Beneficio," a Spanish term for a processing plant at a mine.[15] Downer said of the reduction works:

> As we approach the Hacienda, we are again reminded that it was built for time. The warehouses are of substantial brick structure . . . with furnaces, blacksmith shop, and an open space . . . and occupy a space of several acres. There are, at present, thirteen furnaces in constant operation. They are of solid masonry, stand under cover . . . and are in rows about six feet apart, being forty feet long and eight feet in breadth, and ten in height.[16]

The workers in the furnace yard efficiently processed a variety of grades of ore and produced bottled quicksilver for transport. The yard included storage bins, furnaces and condensers, fuel storage, mercury collection and bottling facilities, and means for removing the spent ore—the tailings. Visitors such as Downer focused on the permanent construction, high technology, and what they described as a continual striving for efficiency in the whole industrial process, and these attributes defined the New Almaden reduction works for the visitor. Whereas the mine was created and operated by the other—that is, people foreign to her: the Mexican and Chilean

FIGURE 1.8 The Patio on the Hill, 1863

The patio, or sorting area, included the long structure in the center of the image and the broad, flat areas surrounding it. It was on the patio that ore was graded and weighed, determining the pay for the groups of contract miners. Discarded rock was dumped to the lower right of the image. (Thanks to Benjamin Johnston for combining the images. Courtesy of the California History Room, images 2010-0642 and 2010-0634 combined.)

FIGURE 1.9 View of the Reduction Works, 1863

The reduction works included many separate furnace systems, built to process the different grades and sizes of ores. Great piles of firewood were stacked in the furnace yard. The sheds to the left of the furnaces sheltered ore ready for the furnaces. The view looks in the opposite direction from Figure 1.5. (Photograph is by Carleton Watkins. Courtesy of the California History Room, 2010-0640.)

miners who worked on contract according to their own systems and understandings—the reduction works was overseen directly by white supervisors and the function and processes were completely intelligible to Downer. Downer found morality in the construction and operation of the reduction works, praising the company in her description of the reduction works for efficiency, productivity, and profitability.

The tale of the great wealth of the Hacienda Nuevo Almaden was where many early travel writers to New Almaden ended their stories. Downer also paid homage to the great value of the mine, but also goes on to praise the owners and managers of the mine in a number of ways. First she compared New Almaden with the 2,000-year-old Almadén Mine in Spain, writing that while the ore at New Almaden was richer, the

Almadén Mine was worked by convict labor, a morally questionable practice. Downer stated that it was through a combination of the rich ore and efficient management that New Almaden could compete in world markets. Downer also went on to credit New Almaden management with a "liberal spirit" in the treatment of the employees, "securing the best men, and retaining them by their best interests."[17]

Interestingly, in the last sentences Downer told significant truths about the business of quicksilver in California. She correctly stated that demand for quicksilver was limited, and that returns for the mining company could be many months in coming due to sales down the West Coast of the Americas and throughout the Pacific. She went on to confess that quicksilver was little used in California at that time and while there were some sales in China and Chile, most of the sales from New Almaden went to the silver mines of Mexico. Each of these truths she related spoke well of the mine owners and managers, for from Downer's account they were not handed wealth, but instead earned it through their knowledge, planning, and superior management of the mine.

As interesting, perhaps, as what Downer said of New Almaden is what she did not say. Most strikingly, she never stated who owned and operated the mine; nor did she give more than the briefest sketch of its history. She did not present herself as a guest of any individual, and she did not introduce the reader to any owner or manager. She never mentioned Barron, Forbes & Co., any British interests in the mine, or the powerful influence on the mine from Mexico. In stating that most of the quicksilver went to Mexico she did not confess that the same company owned the silver mines in Mexico. What Downer did and did not say may have been a limit placed on her for access to the mine—that she not stir up, by means of her article, any more concerns and ill feelings of the British and Mexican ownership of the mine.

An additional element of the New Almaden landscape was the Casa Grande—the grand manager's house, office, and hotel built in 1855 and 1856 at the entrance to the valley—a year or two after Downer's visit (see Figure 1.3). The Casa Grande made a powerful impact on all future visitors to New Almaden for it was one of the grandest homes in California, with the house, gardens, and vineyard presenting a cultured, genteel landscape.[18] Writers detailing a trip to New Almaden almost all recounted meeting the mine manager of the day and commenting on his grand house and gardens. The Casa Grande further strengthened New Almaden as a hacienda—creating a locus for the patron who, in the case of New Almaden as discussed earlier, could be one of many stand-ins for the company.

A visitor during the years of Barron, Forbes & Co. ownership was William Brewer of the Whitney Survey of California. He wrote of his visit to the mine: "A most lovely little town has sprung up by the furnace—neat houses on a long street with a row of fine young shade trees, green yards, pleasant gardens, etc. The superintendent, Mr. Young, is absent. He lives in the most magnificent style, like a prince."[19] Brewer's comment is from his diary, and is less measured than the published writings of most visitors, who praised the gentility and good taste exhibited by the Casa Grande and the hosts.

For example, J. Ross Browne, a writer for *Harpers*, said of a visit to the house:

> A little to the left (upon entering the valley) is seen the capacious mansion ... its massive walls and broad verandas embosomed in shrubbery ... An air of luxury and refinement pervades the premises. All the offices and appointments are in excellent taste, combining simplicity and rural effect with convenience and elegance ... Some of the most delightful days I have spent in California have been in this charming retreat.[20]

The Casa Grande was built when Henry Halleck, who would serve later as President Lincoln's general-in-chief of the Union armies in the Civil War, was the mine manager. Halleck had been California governor Mason's secretary of state and was heavily involved in issues of land claims and landownership in California. He was also a partner in the law firm of Halleck, Peachy, & Billings, which specialized in land claims, and was the main lawyer for the New Almaden Company.[21] Halleck was a strong manager, and oversaw the extensive development of all aspects of New Almaden, including mining, reduction, and the lower village. The Casa Grande was his home at the mine on his twice-monthly visits, and was the center at the mine for the extended group of powerful people interested in the mine. For this group the Casa Grande was a symbol of their wealth and position, and a tool for entertaining well and keeping up appearances for creating and maintaining their status in early California.

The New Almaden Mine landscape was created in the 1840s and 1850s to achieve two ends. First, it was a tool for the production of quicksilver, and in its design and construction the mine landscape provided for the production of quicksilver through the mining, reduction, and bottling of mercury, and for the reproduction of the workforce through the mine camps. Second, the New Almaden Mine landscape created a socially acceptable structure for producing wealth from mercury ore. The often-repeated tropes of Downer and other writers focus on the social acceptability of the mine and its operation—stressing, for example, the morality of the picket fence while overlooking the living conditions of the common laborer. The hacienda was the tool that they knew and that they created in California on the frontier just as they had their Tepic cotton mill in the late 1830s.

Mercury and Power

Barron, Forbes & Co. was able to use mercury in its imperial designs because company officers understood the history of mercury and its use and how they could use mercury to create wealth and power. The techniques they used in the development of New Almaden involved a hybrid system of Spanish and British methods for the control and use of California's mercury as a tool for wealth and power. The development of the Hacienda Nuevo Almaden was the production component of this system, but it was not the whole story. To control mercury meant to control not only its production, but also its trade and use, and the means for this extensive control was refined over centuries of control over the New World by the Spanish Empire.[22]

Throughout the history of the industry, certain attributes of the production, trade, and use of mercury have been used repeatedly across time and place. Some of these attributes derive from the inherent physical qualities of mercury as an element, and others from the ways humans have produced, traded, and used it. These attributes of mercury define a set of structural possibilities that have taken varied concrete forms at different historical moments depending on the dominant political, social, and economic systems of each period. As molten steel always remains steel but can take many forms based on the mold it is poured into, so too have these attributes of mercury dictated the structure of the relationship between mercury and people that had persisted for 400 years (from 1500 to 1900) despite historical contingencies molding it into different forms. How mercury was produced, how it was used, and how it was controlled did not change significantly over these centuries—what changed were the particulars tied to a specific time and place. Despite the political, economic, and cultural shifts that occurred during that period, the attributes of the mercury trade remained true. Each succeeding empire made use of them in a different political and economic context. The attributes serve as a means to compare the role and meanings of mercury across empires and across centuries, and provide a way to understand the mercury mining industry in California in its global historical context. Barron, Forbes & Co. and its successors in the control of California quicksilver molded and modified these attributes of mercury production to their ends.

The first attribute of mercury is that it is rare. Its primary ore, cinnabar, is found in commercially viable concentrations in only a handful of places in the world. One mine, the Almadén Mine in Spain, in continuous production for over 2,000 years, was the source of nearly half of all the mercury produced in the history of the world. Three other mines, the Idrija Mine in Slovenia, the New Almaden Mine in California, and the Huancavelica Mine in Bolivia, produced much of the rest.[23] These four major mines, combined with about ten lesser mines, have produced most of the world's mercury.[24]

Second, mercury has limited uses. Unlike commodities such as gold or grain, mercury is of little use to people in their daily lives, and excepting the occasional mercury thermometer, mercury light switch, compact fluorescent lightbulbs, or mercury medical or dental preparation (all of which consume only small quantities of mercury), mercury is not a consumer good. Nearly all mercury is used commercially.[25] Historically, mercury was the key to refining gold and silver. In Mexico, in 1556, a very efficient means of refining large quantities of gold and silver ore using mercury amalgamation was invented: the "patio process." Although the Romans had used mercury amalgamation to refine gold and silver in small quantities, the patio process greatly increased gold and silver production, creating at times a frantic demand for mercury among mercantilist states developing New World colonies.[26] Until about 1890 and the development of the cyanide process for refining precious metals, mercury was the key to refining bullion. Most important, however, few wanted mercury for its own sake; people mostly wanted what it could be used to create: gold and silver.

Third, mercury was rarely produced where it was used. It has always been a regional, and often a global, market item. With only a handful of major mines in the world, and a larger but still modest number of sites where mercury was used, the geography of the trade in mercury was global, yet nodal: it consisted of a small number of mines, a small number of use sites scattered around the globe, and trading centers and trade routes connecting these sites.

Fourth, because of its rarity, mercury has always been subject to monopoly control. Over time, powerful groups have managed to gain monopoly control of the production, trade, or use of mercury or some combination of these three. With monopoly control, the commodity value of mercury could be increased through regulating production, suppressing the development of other mines, and creating a close relationship between the producers and the commercial users. From the point of view of producers, mercury was a commodity that demanded tight control. As mercury was a necessary commodity in only a few industries, the yearly global demand for the element was largely predictable. The amount used in silver and gold refining and in munitions largely determined yearly demand. As new markets were difficult to develop, the production of mercury was driven by existing demand. However, mercury could generally be produced in quantities that were greater than the yearly demand, and producing more mercury did not mean that you could sell more mercury.

Fifth, the most valuable attribute of mercury was the economic power that could be exercised through the control of its production, trade, and use. The control of mercury has historically contributed to great political and economic power. Mercantilist states such as the Spanish Empire, banking houses such as the Fuggers or Rothschilds, and vertical monopolies such as the one controlled by William Ralston during the Comstock Boom in the American West all used their control over mercury as a tool enabling the control of much more lucrative pursuits. Individuals were made wealthy on mercury production; empires were built and maintained by controlling mercury trade and use. At many times in the history of the industry, the producers and the major consumers of mercury were controlled by the same entities. Whoever managed to control both the production and the commercial use of mercury was on the way to tremendous riches.

Mercury and Empire in the Modern World

Europe achieved world hegemony during the years 1500 to 1900; the production, trade, and use of mercury played a significant role in this achievement. A series of silver and gold mining booms, beginning with the silver boom in central Europe in the late 1400s and early 1500s, continuing to the Central and South American silver and gold mines of the Spanish Empire, and concluding with the gold and silver rushes of the American West, was fueled by the process of mercury amalgamation and the riches that this process produced.

Crucial to understanding the historical significance of mercury is recognizing that it was an important tool in European colonial expansion. European colonial endeavors were based in the extraction of resources from colonies, mining being an obvious example. Especially during the Spanish Empire, colonialism defined the social, political, and economic system in which mercury developed as an important global commodity. For the most part, except in the American West, mercury was controlled by the European core of the world system, and transported to the colonies as a way to both enable and control the production of gold and silver wealth, which was then returned to the European core.[27]

Before 1490 there was one major source of mercury: the Almadén Mine, located halfway between Madrid and Seville.[28] The Spanish Almadén (derived from the Arabic word *al-maʿdin*, meaning "the mine" or "the mineral") had come under Spanish Hapsburg control in the mid-1400s, when the Arabs were pushed out of central Spain.[29] In gaining control of Almadén the Spanish Hapsburgs had an effective monopoly on mercury, and thus on bullion production across Europe, which depended on mercury for amalgamation (although the much more efficient patio process of mercury amalgamation had not yet been invented). At the time the Hapsburgs, both the Spanish and the Austrian branches of the family, were united in a struggle for power in Europe, mostly against the French. Bullion to finance this struggle was in great demand, and the Spanish Hapsburgs used mercury for processing the ore at the silver mines of Guadalcanal, in Spain, and shared their mercury with the Austrian Hapsburgs for use in the silver mines of central Europe.

Although the Hapsburg families were related, the Austrian Hapsburgs wanted their own source of mercury in order to consolidate their economic and political power. In 1490 miners opened the Idrija Mine, about seventy miles northeast of Venice. This and the ancient and immensely rich Almadén Mine formed a two-node geography of mercury in Europe that lasted for over 500 years.[30] With the development of the Idrija Mine, the central European powers (the Austrian Hapsburgs and the Venetian merchants) had a source of mercury that freed them from the Spanish control of the world's mercury supply. In its first days the Idrija Mine was under the control of both Austrian princes and Venetian merchants. Mine companies, under royal appointment, developed and operated different shafts at the Idrija deposit. By arrangement the mining companies at Idrija could only sell to the princes or to Venetian merchants, who then traded throughout Europe and the Mediterranean world. Through this arrangement the Austrian Hapsburgs, together with a few Venetian merchants, were able to both control a large percentage of the global flow of mercury and have a hand in controlling the production of bullion.

Almadén, however, was the richer mine, and was used by the Spanish Empire during its height in the early 1500s as a means of financing its expansion in Europe and the New World. While the New World offered great wealth and opportunity, the Spanish Crown was focused on empire building in Europe; New World wealth was directed to the creation of a Spanish Hapsburg Empire spanning Europe. For the Spanish Empire, Almadén was a bankable asset in an unstable world. The Crown used the value of the Almadén Mine to finance, in part, the ascension of Charles V to Holy Roman emperor in 1519. The Fuggers, a German banking house, were one of the major financial backers of Charles V, and in order to pay back the bankers the Crown negotiated with them the lease of the Maestrazgos, a region of central Spain including Almadén, in 1524. As part of this lease the Fuggers gained control of Almadén, which they held until 1648.[31] Almadén was especially lucrative for the Fuggers because they also controlled the silver mines of Guadalcanal in Spain. With both a mercury mine and silver mines under their control, the Fuggers were able to reap great riches.

The period 1490–1557 encompassed a vast expansion of European powers into the New World, and mercury production and trade were an increasingly important component of this expansion. The silver mines of central Europe were declining by the 1530s, while the mines of the New World were rising and consuming increasingly large supplies of mercury. As a blessing to their colonial plans, Spain controlled the largest mercury mine in the world and was in a position to reap great wealth from the silver mines of the New World, especially after the invention of the patio process. These silver mines consumed all the mercury from Almadén, while mercury from Idrija supplied the rest of the world. For the Spanish a loss of their long-time monopoly on the mercury trade with the development of the Idrija Mine was secondary to the promise of great wealth to be unlocked in the New World by mercury.

Mercury and the Spanish Empire (1557–1820)

The year 1557 marked both the abdication of Charles V and the invention of the patio process of mercury amalgamation, increasing sixfold the precious metal wealth produced in the Spanish Empire. This wealth allowed Spain to rise economically and to claim more power in Europe, despite the misfortunes of Charles V. Before the patio process, the imports of gold and silver from the New World were uneven. From 1492 through the 1530s, the Spanish looted already refined gold and silver from the peoples of the Americas.[32] By 1540, however, there were no more precious metals to loot, and the Spanish invaders sought out and located the mines of silver and gold in the Americas.[33] In 1543 there was a silver rush in Mexico, followed in 1545 by the opening of rich silver mines at Potosí in Upper Peru, which were to become the most important in the New World.[34] For a decade or two, the ore from the Mexican and Peruvian mines was of a chemical composition and richness that allowed the use of established refining techniques, including smelting and small-scale mercury amalgamation.[35] However, by the late 1550s, the mines were deeper and the composition of the ore changed: it was less rich, and the refining techniques were less successful. The patio process of mercury amalgamation was invented in 1557 as a solution to the ore problem, and became a foundation of the system the Spanish colonial administration created to control mercury and thus control the riches of the New World. However, implementing the patio process took years and came too late to help the 1557 bankruptcy of Spain or the abdication of Charles V.

Credit for the invention of the patio process goes to Bartolomé de Medina, a well-to-do wholesale merchant from Seville active in the New World trade. Medina left Seville and his lucrative business interests in his middle age and journeyed to Mexico to pursue methods of refining silver. Through connections with German mining engineers, he had learned basic mercury amalgamation techniques. Medina's goal was to tailor these techniques to both large-scale production and the particular ores of the New World. He was motivated in part by the prospect of royalties and Crown recognition, rewards that he did not receive despite great efforts on his part.[36] The patio process was devised by Medina after years of trial and error and at great personal expense. In his process silver ore was crushed into sand, which was then spread in a thin layer on a patio of stone. Salt and copper sulfate were added and thoroughly mixed in. Mercury was then added and the mix allowed to cure in the heat of the sun for fifteen to forty-five days, being occasionally reworked. The copper sulfate was the key to the process and was Medina's innovation; it was the catalyst for a relatively speedy and complete reaction. As it cured on the patio, the mercury and silver combined to form an amalgam: a pasty substance composed only of mercury and silver. The complete mix from the patio was then washed with water to remove all but the silver amalgam, and this was then heated in a retort, transforming the mercury into a gas while leaving behind pure silver.[37] With Medina's invention, ores of lower quality could be economically refined. As an extra windfall, the massive piles of tailings that were built up around the mines from earlier mining activities could be reworked, extracting silver from ore that the earlier process had missed without further mining expense.[38]

The patio process introduced a new level of industrialization into silver mining that forced changes in the operation of the New World mines. Before the patio process, the Spanish controlled the mines but did not actually work them. Typically Indians took contracts on the mining and refining, paying an agreed-upon quantity and quality of silver to the Spanish and, if successful, keeping any excess silver as profit for themselves. This changed with the patio process, as it needed large-scale capital and a new level of technical knowledge. Large mills, requiring both investment and a new expertise, were established at the mines for the mass processing of ore using Medina's invention.[39] Correspondingly, Indian miners shifted from contractors to supervised labor.

Medina had invented the patio process in Mexico, and the mines there quickly adopted the process. Its adoption was much slower in Peru, where the patio process was not standard practice until the 1570s.[40] The small-scale mine operators saw little benefit to themselves from the process, and saw the process as a danger to their established livelihood. The colonial administration, however, encouraged amalgamation, and soon the new mills produced large quantities of silver at unprecedented low costs, creating a boom in production.

The patio process made mercury both valuable and scarce. The Spanish Hapsburgs were satisfied to keep all the Almadén quicksilver for use within their empire, but soon it was not enough. The wealth that could be unlocked from the New World was directly related to the available supply of mercury, and in the operation of the New World mines mercury was the crucial ingredient that was not supplied locally.[41] Mercury was shipped from the Almadén Mine to the colonies, and at times of short supply deals were made to acquire mercury from Idrija to supplement the Almadén supply. Supplying the New World mines from Europe, however, proved to be both insufficient and too often unpredictable. The Crown needed more mercury, and in 1570 it supported the development of a major mercury mine in Peru called Huancavelica.[42] Huancavelica proved to be a rich mine, and for over 200 years it provided much of the mercury needed in the Peruvian silver mines. With Huancavelica supplying Peru, most of the mercury from Almadén went to the mines of Mexico. With two major mercury mines under its direct control, the Spanish Crown had flexibility in their production and supply; the disruptions in the Almadén supply from wars in Europe were mitigated by mercury from Huancavelica.

The Spanish colonial administration invented a mercury-centered system with four major components to control and regulate bullion production in the New World.[43] The first component was the patio process, allowing silver and gold ore to be efficiently refined. The second component was the control of Almadén and the development of Huancavelica, which supplied the large quantities of mercury required by the patio process. The third component was a series of administrative changes and ordinances instituted in the early 1570s by the new viceroy of Peru, Francisco de Toledo. These changes were created in order to tightly control the production, trade, and use of mercury. The fourth component was the *mita* system of forced Indian labor whereby Spanish landowners had to pay a "mita rent" to the Crown by supplying workers for the mines.

Under Toledo's mercury ordinances, the Spanish Crown forbade the development of any new mercury mines. Then, with total control of the mercury supply, other ordinances secured profits for the Crown at many points in the production, trade, and use of mercury. The Crown taxed the contractors operating the mercury mines, it bought the mercury at a fixed price, and then, as sole sales agent, it charged monopoly prices for the mercury. The Crown used the quantity of mercury sold to the gold and silver miners to calculate the amount of bullion they should be producing, and thus the sale of mercury determined the taxation, or "royal fifth," that was due the Crown as its share of bullion production.[44] Through these ordinances the Spanish Crown manipulated mercury production, trade, and use as primary control mechanisms for silver and gold production in the Americas.

The final piece in the Spanish colonial system for controlling bullion was the mita system. With the vast increase in bullion production came the need for more laborers for Huancavelica and the bullion mines and mills, and in 1572, Toledo instituted the mita system of forced Indian labor.[45] Under this system workers were sent to the mines in relays, called mita, meaning "time" or "turn." This system put tremendous burdens on the Indians, not the least of which was their removal from their home areas and forced relocation to the mines. All mines were dangerous places to work, but the Huancavelica Mine was particular deadly. In the mines the laborers were subject to deadly gases. Cave-ins were all too frequent, and the drastic daily temperature change for the miners—from the hot, deep mines to the frigid air high in the Andes they encountered upon leaving the mines—made pneumonia a common ailment. While underground mining was deadly, the surface work of reducing the cinnabar ore unleashed deadly mercury fumes, and mercury poisoning (also called salivation) was common. While the mita labor system endured for over two centuries, there was a frequent debate among Crown powers concerning the morality of such a system in a Christian civilization.[46] Although these debates reemerged many times during the Spanish colonial years, in the end the mita system persisted.[47]

Once in place, this system of control exercised through the control of mercury secured the wealth of the New World mines for the Spanish Crown. Viceroy Toledo called Huancavelica and the rich silver mines of Potosí the "Pillars of the Empire," and the relationship he created between Huancavelica and Potosí was a spectacular success for the Crown.[48] From 1575 to 1580 there was a 300 percent increase in annual registered silver production in the Potosí District.[49] However, there was resentment against the ordinances that put mercury under Crown control, as expressed by friars in Potosí in 1572, who argued that "to prohibit [free traffic in] mercury and to place it under monopoly is like placing a monopoly on bread or meat, for it is understood that the sustenance of this land depends on the mines of silver, and that they cannot be maintained without mercury."[50]

The changes instituted in the New World in the 1570s—the patio process, the opening of Huancavelica, the mercury ordinances, and the mita system—were the backbone of the colonial mining structure and proved remarkably durable, lasting well over 200 years. It was not until the very end of the eighteenth century that these systems collapsed under a growing fervor of rebellion. Throughout the eighteenth century, the finely interwoven relations of mercury driving the creation of wealth through the production of bullion in the New World started to fray. The New World bullion mines needed more mercury than Almadén and Huancavelica could supply, and to make the problem more acute, by the late eighteenth century the Huancavelica Mine had very low production. As it had during crises in the past, the Spanish Crown again made a contract with the Idrija Mine to purchase its output to supplement the mercury from Almadén to the New World. This relationship lasted from 1785 to 1797.[51]

Napoleon ended this contract when he took control of lands that included the Idrija Mine in 1797, causing a major disruption to the supply of mercury within the Spanish colonies in the New World. With Huancavelica in major decline and Idrija mercury no longer available, the Spanish Crown had to rely fully on Almadén, which alone could not produce the mercury that was demanded. By 1804 Almadén mercury was no longer available at all as Spain was embroiled in the Napoleonic Wars.[52]

Mercury and the British Empire (1820–1863)

In the New World colonies the production of bullion was dependent on mercury, but at the beginning of the nineteenth century there was no supply of mercury in the Spanish colonies. The mercury supplies had disappeared gradually: Idrija mercury was blocked after 1797, Huancavelica had its last good year in 1800, and Almadén mercury was not available after 1804. The lack of mercury in the colonies, combined with a growing revolutionary fervor in the face of the declining Spanish Empire, led to the decline of the New World silver mines early in the nineteenth century.[53] From 1808 to 1814 Napoleon controlled much of Spain, and taking advantage of this fact the colonists in the Americas, particularly the Creoles, looked to better their situations by divorcing themselves from the Spanish. Little wealth came from the New World in these years, and Spain had limited resources with which to control unrest in the colonies. New World forces desiring liberation eventually overwhelmed the old colonial system, and one by one the New World colonies gained independence: Mexico and Peru in 1821, and Bolivia (formed from what was Upper Peru in colonial days) in 1825.[54]

For years after the breakup of the Spanish Empire in the New World and the establishment of independent states, the bullion mines remained in arrested development; only gradually did mining redevelop and a market for mercury reappear. The London house of the Rothschild family filled the mercury supply void by creating a new monopoly on the global trade in mercury by 1835. This house, one of five Rothschild houses in Europe, had gained control of the output of the Almadén Mine by supporting Queen Regent Maria Christina over Don Carlos, in a battle over succession to the Spanish throne. As a result of their support for the Queen Regent, the London Rothschilds received the contract on the Almadén Mine; the Spanish Crown no longer had use for mercury beyond profit from its sale, given the loss of its colonies in the Americas. In the deal the Rothschilds advanced money to the Spanish government against consignments of quicksilver. This contract was immensely valuable to the Rothschilds for the next eighty-five years; in the words of Alphonse Rothschild, Almadén was a "milch cow."[55] As they had in 1524 with the Fuggers, the Spanish powers again used the rich Almadén Mine as a way to secure loans in their unstable country.

The Austrian House of Rothschild had completed the family's mercury monopoly by securing the mercury from Idrija a few years earlier. The monopoly control formed by the collusion of the English and Austrian Rothschilds introduced a new structure into the history of the mercury industry, one in which commercial trading houses, not Crown powers, controlled mercury and could freely trade it where they wished.[56] Operating within the expanding British Empire, the Rothschilds initially made money by being mercury middlemen, selling mercury throughout the world.[57] Following the defeat of Napoleon in 1815, Great Britain emerged as the strongest power in the world. With the ideals of free trade and constitutional government, the British had mechanisms for expanding their economic powers, and quicksilver played an important role in the British development of bullion-producing regions, including southern Africa, Australia, and the Americas. London became the center of the world economy and because of the Rothschilds the center of the mercury trade as well. Soon the Rothschilds managed to leverage the power they had through their control of mercury supplies to exploit the upheavals in the world's silver and gold production and become major mine owners and bullion suppliers themselves.

Barron, Forbes & Co. and the Development of California's Mercury

The breakup of the Spanish Empire in the Americas opened up the region to development by other countries and entities, and British trading houses—riding the rising wave of British global economic dominance—led the way. Barron, Forbes & Co., founded in 1823, grew to be the most important of its kind on the Pacific Coast of the Americas for the next forty years or more.[58] Eustace Barron had been born in Spain of Irish parents, and from one account his manners and demeanor were those of a Spanish gentleman.[59] Barron was also the British consul in Mexico, and his personal combination of both government and business roles served him well for many years.[60] Forbes was Scottish, and unlike Barron was in no way Spanish. In order to run their business, both Barron and Forbes became naturalized citizens of Mexico. In their business and their lives, they combined the rising power of Britain with the reshaping of the former Spanish Empire. The 1820s in the former Spanish domain was a time of instability and rebellion; in this milieu Barron and Forbes were ambitious, opportunistic, adaptable, and immensely successful.

The tool they created that led to their success was their integrated commission house, trading network, and industrial firm centered in Mexico and operating up and down the Pacific Coast of the Americas.[61] While based in the consignment and commission business, Barron, Forbes & Co. also had a sugar plantation, mines, cotton mills, and textile factories. To profit from moving goods and the goods of others, the merchant house had a fleet of ships and railroads. It also owned and speculated in real estate in a number of countries.[62]

Forbes initially had imperial ambitions for California, and these ambitions were detailed in his book published in 1839 entitled *California: A History of Upper and Lower California from Their First Discovery to the Present Time*.[63] Based on the stories of merchants and travelers who had traveled to Alta California, he called for the

British Empire to colonize Upper and Lower California and in the process counter Mexican control and protect what he saw as a land of immense potential from American, Russian, and French designs. The grand plan that Forbes outlined in his book involved Great Britain canceling the large foreign debt owed it by Mexico in exchange for Upper California.[64] This new territory was to be developed and governed by an entity similar to the British East India Company. Forbes's imagination and ambition no doubt included an important role for Barron, Forbes & Co. in the British development of Upper and Lower California. In Forbes's mind the British were the proper people, governed by the proper laws, to make the most of Upper California's potential, potential that would be wasted by Mexicans and their government. He stated that "nothing can be more different from the non-interference with private enterprise, and private conduct, which characterizes the British policy, than the meddling and vexatious interference of the military and civil authorities, which mixes in all the business of life in the present Spanish American countries."[65] Forbes's plans for colonizing Upper California never gained the support of the Foreign Office in London, and he himself, by the early 1840s, realized that California was either going to be independent or taken by the Americans. His focus changed from looking to the British government for development assistance to looking at how to use Barron, Forbes & Co. to gain wealth and power through the development of resources and industry in Mexico and California.

In 1845 word of a quicksilver prospect in Alta California reached Alexander Forbes in Mexico, and by early 1846 Barron, Forbes & Co. had made a deal with the owners of this mine to operate it for a period of sixteen years, during which they would assume all costs involved in developing the mine and would pay a 33 percent royalty to the owners on all quicksilver produced.[66] Barron and Forbes understood that the prosperity of their silver mines was tied to the availability of low-cost quicksilver. The first years, 1846 to 1848, were spent figuring out how to mine and process quicksilver and determining who would manage the mine.[67] To develop the Hacienda Nuevo Almaden, Barron, Forbes & Co. had to import materials and machinery, bring workers from Mexico and Chile, develop and build a production plant, and attract knowledgeable mining engineers.

Barron, Forbes & Co. developed the Hacienda Nuevo Almaden on a frontier that was undergoing revolution. These were years of war and revolution in California and Mexico, encompassing the Mexican War (May 13, 1846, to February 2, 1848), which ended Mexican control in California.[68] From 1846 to 1850, California was literally lawless. For the two years following the war, the United States failed to establish a territorial government, and it was not until Californians ratified a state constitution on November 13, 1849, that the state was on the way to establishing the rule of law.[69]

During these years Barron, Forbes & Co. managed to establish New Almaden as a powerful mining hacienda, doubling world production and backed by an affiliated group of partners who were some of the most powerful men in the state.

By gaining control of New Almaden in the late 1840s and developing it by 1851 to rival the production from the Almadén Mine in Spain, Barron, Forbes & Co. reestablished the mercury systems in Central and South America, and gained perhaps their greatest wealth and power by doing so. Barron, Forbes & Co. managed to reinvent the four components of the mercury system on the California frontier. First, mercury was the most efficient means of refining gold and silver, and thus mercury still commanded a high value for this purpose. For Barron and Forbes, their knowledge of bullion mining made clear to them that if they could secure a source of mercury, then they could control all aspects of gold and silver production at their own mines.[70] Second, Barron, Forbes & Co. controlled the New Almaden Mine, and worked to suppress the development of any other quicksilver prospect. The mine proved to have extremely rich ore, and Barron, Forbes & Co. was able to establish sufficient technology to operate the mine. Third, the fact that Barron, Forbes & Co. was the dominant trading house on the West Coast of the Americas and perhaps in the Pacific meant that the owner-managers could control the trade of mercury. Fourth, they imported Mexican and Chilean miners who had a history of working as wage labor (and perhaps in bondage or peonage) at mines organized like New Almaden, and who were a racial and ethnic group increasingly marginalized in American California.

The Hacienda Nuevo Almaden was the major physical component of the plan of Barron, Forbes & Co. to control the quicksilver industry of California. At the time it gained control of New Almaden, the future of California was up for grabs, with Mexico, Great Britain, and the United States all fighting for control. Thomas O. Larkin saw this struggle, reported it to his superiors, and angled for his own share. The great change the California quicksilver industry embodied, though, was that although mercury had historically been a tool of the state, in California it was the tool of powerful business interests.

In developing Hacienda Nuevo Almaden on the frontier of California, Eustace Barron and Alexander Forbes were doing what they had had done up and down the West Coast of the Americas since the 1820s—move into unstable areas and develop trade and business interests through their merchant house of Barron, Forbes & Co. The organization of their house, like British merchant houses generally, was inherently flexible; there were partnerships with owner-managers who were ready to adapt to ever-changing circumstances.[71] The mediation of risk was central to their business: while trade was inherently risky, Barron, Forbes & Co. also had to contend with near-constant revolution in Mexico and instability up and down the West Coast of the Americas, including in early California. One tool that greatly aided them in dealing with risk—a tool that other merchant houses did not have—was that both men were, at times, British consuls in Mexico, a political position charged with enabling and promoting commerce. This role, paid by the British government while giving them stature in Mexico, undoubtedly advanced their merchant careers.[72] In their business as a merchant house, Barron and Forbes had ships carrying out their trade and real estate interests in ports up and down the West Coast of the Americas. And although for both of them the Barron, Forbes & Co. house was central, each also had his own interests and assets. One or the other partner had interests in cotton mills, textile factories, farming, a sugar plantation, railways, and silver mines.[73]

By the 1850s Barron, Forbes & Co. was the most powerful British house in Mexico.[74] Hacienda Nuevo Almaden was but one piece of its firm, but it was a piece that proved immensely valuable. Readapting the Spanish Empire's mercury-centered system to control and regulate bullion production was the key to the success of Barron, Forbes & Co. and the California mercury industry it invented. It created and developed the New Almaden Mine to tap into the great wealth offered by the cinnabar resource, adapting the mine landscape as a tool to produce wealth. The New Almaden landscape, however, was also created by the struggles of competing groups for getting what they wanted out of the resource. We understand this landscape as a hybrid landscape, created from a long history of people and their relationships over mercury, combined with the particular history of the mine and its influences from Mexico, Spain, and Great Britain, all translated to a rapidly changing California frontier.

Barron, Forbes & Co. used its power and assets to rapidly make New Almaden the most industrialized mine in California, with extensive landholdings, mine works, and settlements, organized on the model of the Mexican mining hacienda. By 1850 New Almaden was an industrialized mine with a wage labor workforce using advanced technology and producing half of the world's mercury—all in the midst of the general insanity of the gold rush. Most telling of the power of Barron and Forbes in early American California is that New Almaden—the land, mines, and settlements—remained a British/Mexican island within developing California until 1863, connected to the rest of the world through the sophisticated trade network of Barron, Forbes & Co.

The attributes of mercury that had shaped the quicksilver industry since the 1500s also shaped the early mercury mining industry in California. However, there were some significant differences between mercury in the American Eempire and those that had come before. Most significantly, in California no Crown or state interests controlled the development of the industry. Instead, multiple owners—Barron, Forbes & Co. and its partners—developed mines, and these people were motivated by money and the power that came with having money, rather than by political or national motives.

Notes

1. California was ceded by Mexico to the United States on February 2, 1848. Gold was discovered January 24, 1848, and accounts of gold in the Sierra foothills appeared in San Francisco newspapers in March 1848.

2. George P. Hammond, *Thomas O. Larkin Papers*, 212.

3. The dispatch was dated October 17, 1845. See James Rawls and Walton Bean, *California*, 80–81.

4. Letters regarding quicksilver are on pages 1, 77, 97, 103, 220, 263, 352, 358. Hammond, *Thomas O. Larkin Papers*.

5. Winks, *Frederick Billings*, 100.

6. Hammond, *Thomas O. Larkin Papers*, 220, 353.

7. The recently appointed military governor, Governor Mason, served from 1847 to 1849.

8. The 200 pounds of mercury produced per day related by Larkin would result in about three flasks (a flask being 76 pounds). To reach the 28,000 flasks produced in 1851, the mine would have to average 75 flasks a day, every day of the year.

9. John Mayo, *Commerce and Contraband on Mexico's West Coast in the Era of Barron, Forbes & Co.* For details on its mills in Tepic, see 370–72.

10. The word *hacienda* is used a number of ways in relation to New Almaden. Commonly the *Hacienda* (uppercase) refers to the lower village and reduction works at New Almaden. *Hacienda* was also used at New Almaden and New Idria to mean the reduction works. I am using the term *hacienda* (lowercase) to refer to the Spanish/Mexican plantation model/system. For information on the hacienda system, see Herman Konrad, *A Jesuit Hacienda in Colonial Mexico*.

11. S. A. Downer, "On Her Trip into the New Almaden Mine," 220–28.

12. Sara Mills, *Discourses of Difference*.

13. Downer, "On Her Trip into the New Almaden Mine," 113.

14. Ibid., 115.

15. Some twentieth-century authors writing about New Almaden also referred to the valley, and sometimes the Casa Grande, as the Hacienda.

16. Downer, "On Her Trip into the New Almaden Mine," 121.

17. Ibid., 122–23.

18. For an interesting look at William E. Barron's (of Bolton, Barron & Co.) own estate and the landscape of the New Almaden Mine, see Peter Coates, "Garden and Mine, Paradise and Purgatory," 147–67.

19. William H. Brewer, *Up and Down California in 1860–1864*, 157. Brewer wrote this entry on August 17, 1861. John Young was a former ship captain for Barron, Forbes & Co. who was made mine superintendent.

20. J. Ross Browne, "Down in the Cinnabar Mines," 550. Browne visited the mine after the American takeover, when Samuel Butterworth was in charge.

21. Winks, *Frederick Billings*, 102. The architect for the Casa Grande was Gordon Cummings, who was also the architect for the famous Montgomery Block in San Francisco, a building developed by Halleck, Peachy and Billings.

22. Nicholas Robins, *Mercury, Mining, and Empire*, chap. 1.

23. The Idrija Mine has been part of many kingdoms, empires, and nations during its long history. For most of the time under discussion here, Idrija was part of Austria. It is currently in Slovenia.

24. Other mines include Monte Amiata in Italy; Guadalcázar in Mexico; the mines of Kwei-Chan in China; mines at Nikitowka in Russia; the mines of New Idria, Oat Hill, Great Western, and Knoxville in California; and the Chisos Mine in Texas.

25. Historically mercury has been used for amalgamation and for mercury fulminate used in bomb detonators. Amalgamation is the process of creating an alloy of mercury with another metal. Mercury readily combines with gold and silver, aiding in the refining of these metals. Mercury is also used in the manufacturing of paper and paints, in medical and chemical preparations, and in electronics.

26. Accumulating bullion was a goal of many early modern European states, part of mercantilism, an economic policy of state management of the economy to enhance state power.

27. Janet Abu-Lughod, *Before European Hegemony*. She argues that "world systems do not rise and fall in the same way that nations, empires, or civilizations do. Rather, they rise when integration increases and they decline when connections along older pathways decay" (367).

28. There was probably an important source of mercury in China at this time. However, the Almadén Mine was the source of mercury for traders who were active in the Mediterranean and Atlantic worlds. See Fernand Braudel, *The Mediterranean and the Mediterranean World in the Age of Philip II*.

29. The Spanish branch of the Hapsburgs had united Castile and Aragon in 1479 and had pushed Islamic rule out of Spain with the fall of Granada in 1492.

30. Mining activities ceased at Almadén in 2002, while mining at Idrija ended in the early 1990s, ending the 500-year reign of Almadén and Idrija.

31. In this arrangement, the proceeds due the Crown from the Maestrazgos were leased in return for an advance from the Fuggers. From 1563 to 1645 the Fuggers had full proprietorship of Almadén—they brought in advanced German mining techniques and introduced new reduction technology.

32. In 1518 the Aztec Empire was crushed, and by the late 1530s the Inca Empire was at its end.

33. Silver was the most important precious metal due to the vast quantities found.

34. This silver rush led to the development of major mines, including Zacatecas, Guanajuato, and the Real del Monte at Pachuca. Otis E. Young, *Western Mining*, 59.

35. Smelting is a process whereby the ore is melted or fused, producing a chemical change separating the metal.

36. For a thorough recounting of the life of Bartolomé de Medina see Alan Probert, "Bartolomé de Medina."

37. A quicksilver retort is a furnace type that keeps the products of combustion (such as smoke from a wood fire) separate from the amalgam. A simple quicksilver retort was an iron pipe into which the mercury/bullion amalgam was placed and the ends sealed. When heated, the components of the amalgam separated in the pipe while remaining uncontaminated by the by-products of the heat source.

38. The term *tailings* refers to ore that has been processed.

39. See Peter J. Bakewell, *Mines of Silver and Gold in the Americas*, for a detailed discussion of changes caused by the introduction of the patio process, esp. 93–95.

40. Mexico and Peru were the two richest mining centers in the Spanish colonies, and the mining districts had significant differences. Peru was geographically isolated, making communication difficult and increasing the cost of transportation. Reaching Peru by sea meant offloading ships in Panama, transporting materials by land over the isthmus, and then reshipping down the Pacific Coast of South America. Ships could also sail around South America to reach Peru. Despite these difficulties, the mines of Peru were developed earlier and until around 1700 were the most productive mines in the colonies. The mines of Mexico were more accessible, had higher-grade ores, and had lower operational costs. Yet, because the Potosí mines in Peru were so rich and had attracted great attention, the Mexican mines did not become the dominant producers until after 1700.

41. The Consejo de Indias and the Consejo de Hacienda in Castile were the two bodies that, among their many other responsibilities, were responsible for the production, transport, and distribution of mercury.

42. The Crown established an ordinance against the development of other mercury mines, ensuring that minor deposits would not interfere with Crown control of the metal.

43. These ideas on the four components are derived from the work of historian Peter Bakewell.

44. The other power in the mining districts, the Miner's Guild, was responsible for operating the mines. The Crown exercised control over the guild in the following way: it advanced the guild a sum of money and a supply of mercury, it provided a fixed number of laborers under the mita system, and then it expected bullion to be returned to it, the quantity being based on the supply of mercury that had been allotted.

45. See chap. 2 in Arthur Preston Whitaker, *The Huancavelica Mercury Mine*.

46. Many authors have written on the mita system, including Peter J. Bakewell, *Miners of the Red Mountain*; Bakewell, *Mines of Silver and Gold in the Americas*; and Whitaker, *The Huancavelica Mercury Mine*. Over the years, the system was mitigated in a number of ways. Most mines had a double labor structure of both mita labor and wage laborers, although in practice the difference in the degree of coercion imposed on different groups of laborers was slight. Wage laborers often handled the more skilled tasks, whereas the mita laborers—

because they were cycled through the mines—performed unskilled tasks. For the *corregidores* (Spanish landowners), who were under obligation to provide mita rent, a cash payment to the Crown was also accepted in some situations.

47. The other major mercury mines each had different labor situations. At Almadén the labor force included slaves and convicted criminals. These groups labored in the mines from the early 1500s to about 1750. Workers were kept in chains, and infantry was stationed nearby. The laborers generally worked one month underground followed by two months laboring aboveground. At Idrija the workers tended to be a working class and not an underclass, and workers were reputedly rotated among jobs in order to limit the injurious effects of any one type of work.

48. For more on the interrelationship of Huancavelica and Potosí, see Gwendolin B. Cobb, "Potosí and Huancavelica."

49. Bakewell, *Miners of the Red Mountain*.

50. Peter J. Bakewell, *Silver Mining and Society in Colonial Mexico*, 150. This quotation is the friars' reaction to mercury being put under Crown control in New Spain by Viceroy Enríquez in 1572. Viceroy Toledo instituted similar ordinances in Peru.

51. The Idrija Mine increased production during this period, although at the expense of working through all its available ore, creating dire circumstances for the mine in the early nineteenth century.

52. The control of the shipping of mercury had been a problem for the Spanish Empire from the early sixteenth century. Pirates, navies of other nations, and smugglers all had to be controlled by the Empire.

53. By 1810 the Huancavelica Mine was barely producing any mercury, and by 1816 it was closed.

54. Revolutions were not confined to Europe and the Americas. During the nineteenth century, China experienced a series of wars and also an internal rebellion. The Opium Wars, fought against Great Britain from 1839 to 1842, resulted in the opening of new port cities in China and freer trade with the West. The Taiping Rebellion of 1850 to 1873 furthered the opening of China to the West. The Chinese had been major consumers of mercury for hundreds of years, primarily for the production of vermillion, a red pigment. The Spanish had established an Acapulco-Manila trade, including mercury, in the late 1500s, connecting their New World empire with ports in the Canton area.

55. Niall Ferguson, *The House of Rothschild*, 168.

56. The Fuggers were a commercial trading house of sorts, but by agreement they had to sell their mercury within the Spanish Empire. The Rothschilds continued their control of mercury well into the twentieth century, and Almadén in particular proved to be both a reliable source of income and a crucial piece of their mining empire.

57. For an interesting study of the changes in mercury transport in this era, see Tristan Platt, "Container Transport," 205–54.

58. Eustace Barron (b. 1790) and Alexander Forbes (b. 1778) were original partners in the firm. For the history of Barron, Forbes & Co., see Mayo, *Commerce and Contraband on Mexico's West Coast in the Era of Barron, Forbes & Co.*

59. Frederick Billings, *Letters from Mexico*.

60. The great extent of Eustace Barron's wealth and power in Mexico is evident in ibid. On July 12, 1876, the *New York Times*, in the article "Society in Mexico—A Grand Ball at the Mexican Capital, the Festival Given by Its Richest Resident—An Enchanting Scene" details a society event hosted by William Barron, heir of Eustace Barron.

61. Alexander Forbes, *California*.

62. For the range of Barron, Forbes & Co. business interests see Hilarie J. Heath, "British Merchant Houses in Mexico, 1821–1860," 287.

63. Forbes, *California*.

64. Ibid. See the introduction by Herbert Ingram Priestley, xviii–xix. Priestley refers to contact between Alexander Forbes and Robert Crichton Wyllie, of the London Council of the Holders of the Bonds of Mexico, where they developed the plan that Forbes proposed. In the early 1840s, under U.S. president John Tyler and his secretary of state, Daniel Webster, there was an American proposal for the United States to cancel Mexico's debt in exchange for California.

65. Ibid., 201.

66. Following Mexican mining tradition, the ownership of the mine had been divided into twenty-four shares. By the end of 1847, Barron, Forbes & Co. not only had the mine under contract, but Alexander Forbes had managed to purchase eleven of the twenty-four shares. Four of the shares had come from Andrés Castillero, the midlevel Mexican official who had claimed the mine after identifying the ore as cinnabar. In selling shares to Forbes, Castillero was keeping the mine out of American hands by placing it firmly in the hands of the powerful (and British) Barron, Forbes & Co. Alexander Forbes also purchased the four shares owned by General José Castro, the Mexican military commander of California. The final three shares came to Forbes from Padre Real, of the Mission Santa Clara, who by Mexican law could not own shares in a mine as a padre. Forbes, *California*. See the introduction by Herbert Ingram Priestley, xiv–xv. See also Jimmie Schneider, *Quicksilver*, 16–18. Castillero had originally divided the 24 shares as follows: 12 shares for himself, 4 to General Castro, 4 to Padre Real, and 4 to the Robles brothers. Tracing the subsequent history of each of these shares has been a minor historical industry. See the official court record: Castillero and United States District Court (California: Northern District), *The United States v. Andrés Castillero*.

67. During these years, there was small-scale production at the mine. See Coomes, "From Pooyi to the New Almaden Mercury Mine," 70–80, for a discussion of the early development of the mine and the Native Americans who worked at the mine.

68. James J. Rawls and Walton Bean, *California*; Neal Harlow, *California Conquered*.

69. California was admitted to the Union on September 9, 1850.

70. The Mexican government recognized the power a Mexican mercury mine would provide the country by lessening the reliance on European quicksilver, and in 1821 the ruling Sovereign Provincial Committee of Mexico had facilitated mercury exploration by easing the rules governing prospecting and developing mines. In 1842, powers in Mexico established the Board of Encouragement and Improvement of the Mining Corporation, which actively

promoted mercury exploration, offering substantial rewards for the discovery and development of quicksilver mines. Andrés Castillero and United States District Court (California: Northern District), *The United States v. Andrés Castillero*; Mary L. Coomes, "From Pooyi to the New Almaden Mercury Mine," 52–53.

71. Hilarie Heath, "British Merchant Houses in Mexico, 1821–1869," 268.

72. Mayo, *Commerce and Contraband on Mexico's West Coast in the Era of Barron, Forbes & Co., 1821–1859*, 68–70. Eustace Barron was vice-consul and consul from 1824 to 1848; Alexander Forbes also served from time to time.

73. Heath, "British Merchant Houses in Mexico, 1821–1860," 287.

74. Mayo, *Commerce and Contraband on Mexico's West Coast in the Era of Barron, Forbes & Co., 1821–1859*, 392. Another powerful British merchant house, Jecker, Torre & Co., was a part owner of the New Almaden Mine from 1849 to 1852, when Barron, Forbes & Co. bought its interest.

The quicksilver industry in California was a capitalist one; capitalists used mercury to make money, and this money gave them power. Since the sixteenth century, mercury had been a tool used by states to control bullion production. Mercury was used by states as a means to achieve their mercantilist goals; mercury allowed them to accumulate bullion, and state mercury monopolies were crucial to these policies. In California, however, mercury was not a state-controlled political commodity in the same way. It was, instead, a financial commodity, although still used by capitalists to control bullion production. California quicksilver was worth the money and influence it could be exchanged for.

2

Money and Power in the California Mercury Landscape

It takes great wealth to produce great wealth in California.
—Carey McWilliams, *California: The Great Exception*, 36

The 1850s: Doing Business in Early California

From the 1840s until 1863, Barron, Forbes & Co. and its partners controlled the production, and much of the trade and use, of quicksilver in the Pacific world. They managed to maintain their powerful hold on the industry by suppressing the development of new mines, underpricing quicksilver from Europe, and, most important, building business relationships in California and throughout the Pacific. These relationships, originally negotiated around mercury's role as an industrial tool of empire, transformed into relationships between powerful capitalists interested in wide-ranging aspects of the development of California and the Pacific Coast of the Americas. These men understood that the control of mercury was an important tool for acquiring the wealth and power they desired.

To manage the affairs of the mine in California, Barron, Forbes & Co. helped establish the firm of Bolton, Barron & Co. in the early 1850s.[1] Two younger men ran this office: William E. Barron (d. 1871), a son or nephew of Eustace Barron, and James Robert Bolton (1817–90), a former American consul in Mazatlán who was involved in mine affairs early on for his wealth and connections.[2] Thomas Bell (1819–92), a Barron, Forbes & Co. employee eventually known as the "Quicksilver King of the West," became

DOI: 10.5876/9781607322436:c02

involved in the quicksilver trade in the 1850s through working for Bolton, Barron & Co. The company these men developed was part of the Barron, Forbes & Co. network, handled all quicksilver sales from New Almaden, and was the San Francisco office of the mine.[3] Working out of San Francisco, Bolton, Barron, and Bell had the responsibility and the opportunity to use their control of the marketing and sales of New Almaden mercury to build their power in California and beyond. Through the association with Barron, Forbes & Co., these men inherited a strong reputation that aided them in their business dealings throughout the Pacific and in securing favorable rates of interest, essential given the six months to two years that mercury often took to sell in far-flung points in the Pacific Basin.[4]

Bolton, Barron, and Bell rapidly transformed the British imperial business model of Barron, Forbes & Co. into a business model suited to the demands of San Francisco, California, and the United States. They focused on New Almaden mercury as their primary asset, developing their trading and commission house with the sale of quicksilver as their primary business. International sales through the Barron, Forbes & Co. network were their bread and butter. In the early 1850s Bolton, Barron & Co. formed a partnership with the Rothschilds to control world mercury markets. Their account book shows that by December 1850, the company had negotiated a contract with the Rothschilds whereby two-thirds of all sales in Mexico and South America went to the "Negotiation de Nuevo de Almaden" and one-third of all sales went to L. Davidson, the Rothschilds' agent.[5] This combination, however, fell apart after about six months, and by October 1851, both parties were selling flasks of mercury for about $85, when they had previously been sold for well over $100. For May 1, 1852, the Bolton, Barron & Co. account book reads, "We have been compelled to make a considerable reduction of the price in our competition with the agent of Messrs Rothschilds in Mexico and South America."[6] By the third statement, covering most of the sales in 1852, the average price of a flask of quicksilver had declined to thirty-five dollars.[7] Bolton, Barron & Co. and Barron, Forbes & Co. understood from the beginning that their competition with the Rothschilds would result in large price declines, although as the records show, they were willing at first to make a deal with the Rothschilds to maintain prices. However, as the New Almaden Mine was further developed in 1851 and 1852—and as the mine produced extremely rich ore—the New Almaden Company realized that they could still make an excellent profit despite the decline in price, and they did battle with the Rothschilds for control of the global mercury markets (Figure 2.1).

Although the European mines had the best access to the Atlantic and Mediterranean markets, the New Almaden Mine had the best access to Pacific markets, and Bolton, Barron & Co. focused on the distribution and sale of the mercury throughout the

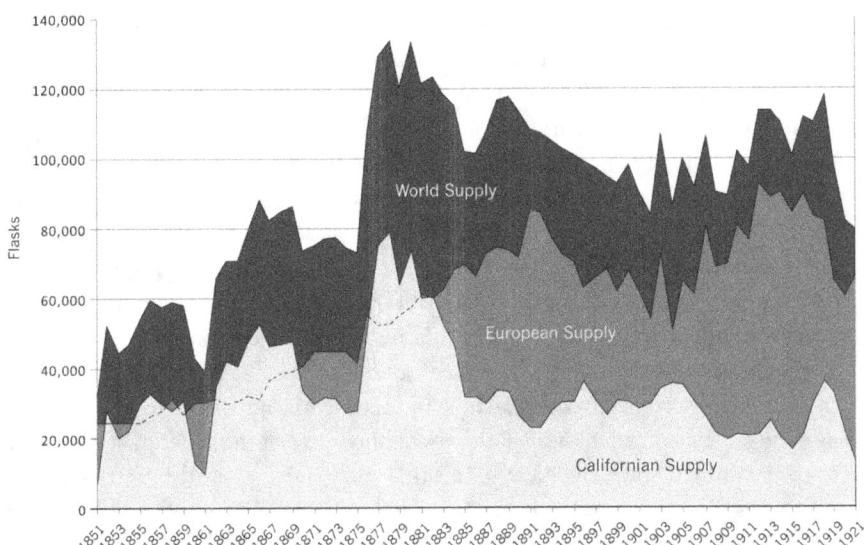

FIGURE 2.1 Comparative World Mercury Production, 1851–1921
From 1851 until the mid-1880s, California mercury production fluctuated wildly while European production (from Almadén and Idrija) steadily increased. The erratic supply of mercury from California wreaked havoc on the stability of prices and supplies on world markets during these years. (Production figures are compiled from David T. Day, *Mineral Resources of the United States*, an annual publication.)

Pacific, making the most of their geographic advantage. Using 1854 as an example: 30,004 flasks were produced at the New Almaden Mine, of which 20,963 flasks, about 70 percent of the total, were exported.[8] Shipments to China, Mexico, and countries in South America accounted for almost all of the exports. Bolton, Barron & Co. shipped mercury to prominent merchant houses in cities around the Pacific, and most of these shipments were on consignment, although there were also direct orders. These far-flung houses then made sales based on preexisting arrangements. At any time Bolton, Barron & Co. had thousands of flasks of quicksilver stored in warehouses in cities around the Pacific, and these flasks could be sold, held, or moved as market conditions or their competition with the Rothschilds warranted. The time it took for a flask of mercury bottled at New Almaden to be shipped to Hong Kong, to be sold, and the money received in San Francisco—perhaps via London—was as long as two years.

For most of the 1850s the New Almaden Mine was the only mercury mine in the Pacific, giving Barron, Forbes & Co.; Bolton, Barron & Co.; and the minor owners of the mine a tremendous geographical advantage over the Rothschilds on mercury sales to the Pacific Rim. Barron, Forbes & Co. profited the most, both as majority owner and from the use of the mercury at their silver mines. Because of their powerful hold over Pacific mercury supplies they could charge other silver mines a premium for mercury. The minor investors in New Almaden made simple profits from mercury sales, and also made business partnerships based on their connection to New Almaden. Bolton, Barron & Co., however, were the long-term winners. Although early on they were agents selling mercury for a percentage of the profit, in the long term they built business connections in California that gave them opportunities to prosper as conditions changed in California and in the mercury industry.

Throughout the 1850s Barron, Forbes & Co. and Bolton, Barron & Co. effectively kept a monopoly on mercury production and trade and kept a hand in mercury use: they owned New Almaden; in 1858 they bought control of the other major quicksilver prospect in the state, the New Idria Mine; they purchased other mercury prospects throughout the state; and they invested in other California industries. Most important, while in many ways the New Almaden Mine remained a British and Spanish island of influence in California, the young businessmen of Bolton, Barron & Co. were thriving in the rapidly transforming business center of San Francisco, and in the process capitalizing on the connections and opportunities their control of mercury gave them.

The 1860s: Land Battles over the New Almaden Mine

The force that wrested ownership of New Almaden from Barron, Forbes & Co., bringing major changes to the mercury industry, was legal battles over the ownership of land in California. These battles began at least as early as the end of the Mexican-American War, in 1848, and mostly involved the validity of Mexican land grants under American law. Battles over the ownership of the New Almaden and New Idria Mines were prominent and well publicized at the time. The battle over the New Almaden Mine, which Barron, Forbes & Co. fought for over fifteen years until the company eventually sold the mine and walked away from court in 1863, was the most prominent land case in the history of early California, culminating in a landmark U.S. Supreme Court decision. The New Idria Mine land case, which the partnered interests of Barron, Forbes & Co. won, also reached the U.S. Supreme Court and took nearly twenty years to work its way through the courts.[9] These land case battles and eventual resolutions, at the forefront of the important issue of land ownership in early California, transformed the mercury mining industry through demonstrating that mercury production was only part of the key to using mercury to unlock great power and wealth in California and the West. Having mercury was of limited use if one had no power to trade or use it.

In the case of the New Almaden landownership battle, American interests—backed by East Coast capital—managed to wrest control of the mine from the British/Mexican ownership of Barron, Forbes & Co. during the upheaval of the American Civil War. The eventual result was that a new company formed by East Coast investors, the Quicksilver Mining Company (QMC), took over control of the New Almaden Mine in California in November 1863. But in a brilliant business strategy enacted over many years, the firm of Barron & Co. (renamed after Bolton left) retained for company members control of significant quicksilver production from New Idria, and nearly all of the trade in mercury and control of the use of mercury in mines and mills that it or its associates controlled. This control of mercury enabled Barron & Co. and its partners to amass great wealth and power, whereas the new American owners of the New Almaden Mine made comparably minor profits from simply producing mercury. Barron & Co. managed to keep control of mercury production in the state by buying the New Idria Mine; continuing to control the Pacific trade in mercury through the networks of Barron, Forbes & Co.; and, most important, entering partnerships in a range of business interests with the most powerful banking and mining men in the West, men who were able to use mercury in building their fortunes.[10]

The New Almaden land claims cases and the legal battles over the New Idria Mine were highly publicized. For years the major national papers covered the cases, and California papers featured every twist and turn. Opinion pamphlets were published supporting many points of view. The cases were much discussed and followed with intense interest because decisions in the cases would bear heavily on questions of landownership and mining rights throughout California. One early traveler, William Brewer—a member of the Whitney Survey—expressed his opinion in a diary entry written at New Almaden on August 16, 1861:

> No wonder there has been such legal knavery to get this mine, when we consider its value. Every rich mine is claimed by some ranch owner. These old Spanish grants were in the valleys; and, when a mine is discovered, an attempt is made to float the claim to the hills. Two separate ranches, miles apart and miles from the mine, have claimed it, and immense sums expended to get possession. The company (Barron, Forbes & Co.) has spent nearly a million dollars in defending its claim—over half a million has been spent in lawyers' fees alone, I hear. The same at New Idria—it was claimed by a ranch, the *nearest edge of which is fifteen miles off*. And this is only a sample of the way such things go here. Were I with you I could relate schemes of deeply laid fraud, villainy, rascality, perjury, and wickedness in land titles that would stagger your belief, yet strictly

true. The uncertainty of land titles and the Spanish-grant system are doing more than all other causes combined to retard the healthy growth of this state.[11]

Barron, Forbes & Co. and their partners in the mine (referred to collectively in the legal proceedings as the New Almaden Company) based their ownership on the mine claim and land grants of Andrés Castillero, a Mexican official who had recognized the distinct cinnabar ore in 1845 and proceeded to file claim to the mine. The problem with Castillero's claim was that in comparison to the American system of land survey and boundary determination in use after California statehood, the Mexican system that was in place prior to the American conquest (the system used by Castillero) was vague: boundaries were not precisely measured, nor were the bureaucratic steps necessary to claim land clearly defined. This imprecision by American legal standards meant that it was not possible for any party to prove ownership of the New Almaden Mine to the American standard, based on the existing information from the Mexican system. The American and Mexican systems could not be reconciled. This led to the ownership of the New Almaden Mine being a matter of interpretation, and thus to a social, cultural, and legal battle that was waged at great expense in both the law courts and the courts of public opinion.[12]

The New Almaden company had dealt with land issues from the beginning of their involvement with New Almaden, and losing the mine due to these issues was always a possibility. The eventual loss of the mine became a high probability when in January 1856, the California Land Commission upheld the claim to the mine but denied Castillero's colonist grant, effectively limiting the New Almaden Company's claim to a small area right around the original mine entrance.[13] With this ruling, parties with designs on the mine sensed vulnerability in the Castillero Claim and organized major efforts to gain control of the mine. These forces raised enough doubt about the ownership of the mine to get a court injunction barring the New Almaden Company from mining ore from October 1858 until January 1861. During these years Barron, Forbes & Co. and Barron & Co. began to see that losing the mine was immanent, and although they continued the legal battle, they also planned for the possibility of losing the mine by flooding world markets with quicksilver, buying other mines and prospects throughout the state, and solidifying business partnerships based in the production, trade, and use of mercury.

In November 1863 the New Almaden Mine became the property of the QMC, a large and complexly associated group of East Coast investors. Samuel Butterworth, a lawyer and opportunist with a multifaceted past, was the front man who helped combine one group of claimants that he had brought together—named the Quicksilver Mining Company—with another group also laying claim to the mine, the California Quicksilver Association.[14] By the spring of 1861 the combined groups, adopting the

Quicksilver Mining Company name, were well funded (via a $10 million initial stock offering) and poised to do battle for the New Almaden Mine. During the years the mine was closed by injunction, the QMC made an offer to the New Almaden Company to buy the mine, but in the spring of 1861 the court lifted the injunction on the mine, and the New Almaden Company decided to continue the fight in the Supreme Court.

Due to the Civil War, the Supreme Court did not hear the case until January 30, 1863, two years after the appeals had been filed.[15] During these years the New Almaden Company worked the mine extremely hard and produced unprecedented quantities of quicksilver that they used to flood world markets. The QMC, meanwhile, used its vast capital to "mine Washington" for support and further backing, while its stock was subject to wild speculation.[16] The fate of the New Almaden Mine was an important issue in Washington; the North wanted New Almaden secure from Southern interests, and preferably in the hands of Northern businessmen, not in the hands of British/Mexican interests. New Almaden quicksilver was valuable for bullion production and for munitions production, both central concerns in wartime.[17]

The decision, rendered on March 10, 1863, was a 4 to 3 vote in which the majority found that the Castillero Mine Claim (the one claim that previous courts had upheld) was invalid because Castillero had not followed the Mexican laws precisely in claiming the mine. The court also rejected his colonist grant. The New Almaden Company still had one hope after this decision, that the mine would be found to be on the Berryessa Grant (a grant they had bought from a local landowning family) and not on the Larios Grant (which the QMC owned). This issue was before the Supreme Court in what was known as the Fossatt case.[18]

The powerful forces of the QMC continued to work behind the scenes after the Supreme Court decision in March. The actual possession of the mine worried these powers—as the old saying goes, possession is nine-tenths of the law, and members of the QMC did not want to take any chances with the Fossatt case. Again exercising their power in Washington, the QMC directed Leonard Swetl—a former law partner of President Lincoln who had large economic interests in the QMC—to have the president sign a writ for the government to take over the mine.[19] Lincoln signed the writ on May 8, 1863, and soon after the signing Swetl traveled west with Samuel Butterworth; Butterworth planned to take over the mine once the government took possession.[20] Armed troops were sent from a garrison at Benicia, California, to the mine, where they were met by armed men from the mine, resulting in days of standoff. In the meantime, figures for both the New Almaden Company and the QMC lobbied in Washington. Frederick Billings, a lawyer for the New Almaden Company, sent a telegram to Henry Halleck—his law partner, the former manager of the New

Almaden Mine, and Lincoln's general-in-chief—warning, "There is great excitement, and unless the mandate is revoked the State [California] is in danger of being lost to the Union."[21] The writ signed by Lincoln raised great concern throughout California and the West because most mines in the West could be subject to this writ in the same way that the New Almaden Mine was. Precious metal mills throughout the West were also concerned that the government would sue for the quicksilver that they had used, claiming that they had gained it fraudulently. Lincoln was caught between the powerful QMC interests and the fear of civil unrest in the West brought on by this writ.

By midsummer 1863 Lincoln retracted the writ, ending the armed standoff at the mine and easing the unrest throughout the mining regions of the West. However, by the end of the summer, the New Almaden Company sold the mine to the QMC for $1.75 million, much less than most newspaper articles reported it to be worth. The Fossatt case had not yet been decided when the sale took place. The Supreme Court, however, decided the case in the spring of 1864, finding that the mine was on the Larios Grant. Hubert Howe Bancroft's staff interviewed Thomas Bell in 1886 concerning the sale of the New Almaden Mine to the Quicksilver Mining Company. Bell claimed simply that the New Almaden Company believed the mine to be in jeopardy, and that it sold it for what it thought it was worth.[22] At this time, 1886, Bell was an old man, and most of his former associates in the New Almaden Company were long dead. Bell's answer to Bancroft's question was diplomatic, raising no ghosts. What is hidden by Bell's answer, however, is that he was one of the big winners as the result of all the litigation surrounding the New Almaden Mine. With William Barron, he eventually inherited the Barron, Forbes & Co. contracts for selling quicksilver, he and his associates owned partial interest in the New Idria Mine, and he had established partnerships with the most powerful business interests in California. He and William Barron shared the wealth until Barron's death in 1871, after which Bell became, in the words of H. H. Bancroft, the undisputed "Quicksilver King of the Pacific Coast," the heir to the Barron, Forbes & Co. legacy.[23]

Financial success, however, eluded the Quicksilver Mining Company once it had ownership of New Almaden. As soon as the QMC took over the mine in November, 1863, it faced daunting costs, including paying the New Almaden Company for the mine, providing operational costs for the day-to-day working of the mine, and financing the costs of selling mercury on the world markets, where returns from sales could be six months and often longer in reaching San Francisco or New York. The management of these financial challenges fell to Samuel Butterworth, who left the presidency of the QMC in New York and took over as general agent and factor at the mine on July 1, 1864.[24] Butterworth's background included being U.S. district attorney for Mississippi and superintendent of the U.S. Army under President Buchanan. In the late 1850s, when he was organizing the various groups attacking the New Almaden Company, Butterworth was in New York City and was affiliated with Tammany Hall politics. For his services in California the QMC paid him a tremendous salary for the time: $15,000 a year, plus $6,000 for house expenses. The Casa Grande was renovated for him, although Butterworth spent a great deal of time in San Francisco, where he was active in social life.

While Butterworth was making good money as general agent and factor for the mine, he may also have believed—at least in part—some of the arguments he and his fellow organizers of the QMC had made to attract investors, and that he would gain more wealth from the truth of these arguments. An 1858 pamphlet shopping the takeover scheme of the New Almaden Mine to East Coast investors argued the true value of the mine:

> These mines (the world's great quicksilver mines) are not only desirable on account of their great pecuniary value, but on account of the influence they must exercise, not only on the productions of gold and silver, but also upon the whole commerce of the country... The party who controls these mines must have great influence on the markets for bullion, and to some extent on the gold and silver coin of the world.[25]

In this pamphlet the author, James Eldridge, made two major assertions regarding his organization: that they were the owners of the New Almaden Mine (although their property had yet to be transferred by the courts) and that the organization was a collection of merchants and bankers who did business in, and had connections in, the major mercury markets of the world and that they could effectively deal in mercury. The first of these assertions proved to be true in practice: the courts eventually awarded the New Almaden Mine to the QMC, which Eldridge's group had joined. The second assertion, however, was false, and this prevented the QMC from ever reaping more than the simple production value of the mercury from the mine.

Although doomed to failure, between July 1864 and April 1866, Butterworth struggled to sell New Almaden mercury on the world markets. His attempts and subsequent failures are detailed in the correspondence copybooks from company records.[26] He was faced with two major problems. First, he had to establish new marketing and banking arrangements with selling agents and banking houses around the world, while directly competing with Barron & Co. and the Spanish government, both of which had warehouses of mercury in markets around the world.[27] Second, he was faced with selling unprecedented quantities of mercury—1864 and 1865 were the two years of greatest production in the history of the New Almaden Mine. Once the QMC took over the mine, the company was faced with tremendous costs both from purchasing the mine and from legal fees. In an effort to pay these costs, the directors pushed the mine superintendent to produce as much quicksilver as possible. In 1864, 42,489 flasks of mercury were produced, and in 1865, 47,194 were produced, the largest production of any year (Figure 2.2).[28]

In addition to selling mercury, Butterworth had responsibility for the day-to-day operations of the mine. His goal was to exact major changes in mine operation and in the relationship between the company and the workers. He inherited the hacienda system established by Barron, Forbes & Co., and although this system had changed and developed over the years, it was to Butterworth a relic of another time and culture that demanded change. Although Butterworth was trying to sell mercury, he was also radically transforming the mine and mining camp environments, a process that was full of strife.

The cost of purchasing the mine, the low price of quicksilver, and the high cost of developing the mine and transforming its operation all added up quickly to a financial crisis for Butterworth and the QMC. In the 1864 Annual Report of the Quicksilver Mining Company, the president, William Bond, stated once more the dream of the QMC for the New Almaden Mine: "The influence and power exerted by the proprietors and lessees of that mine [Almadén in Spain], before the opening of the California mine, in controlling the bullion of the world by virtue of the monopoly of quicksilver, is a well-known fact in history, and adds a dignity and national importance as well as intrinsic value to the ownership of these mines."[29] Unfortunately for the QMC shareholders, the company never realized this dream. The problem for the QMC was how to create or join a monopoly on quicksilver. During 1864 and 1865 the New Almaden Mine produced as much mercury as the Almadén Mine in Spain. This fact raised the expectations of the QMC (they owned a mine that rivaled the richest quicksilver mine in the world) while doubling the world's supply of mercury and reducing its price. The New Almaden Company had developed its production and sales together, and always regulated the supply. The QMC was not able to regulate the supply; it was producing too much mercury and had insufficient means of selling it. Butterworth made a number of attempts to drive the Spanish government's Almaden mercury from major markets by flooding these markets with surplus New Almaden mercury. The 1864 Annual Report states that 18,908 flasks of mercury were shipped to China, up from 8,889 flasks in 1863. This strategy did have the long-term effect of establishing a secure market for New Almaden mercury in China through the 1870s. However, in the short term, the QMC made little money on sales in China.[30] Also, in an 1864 letter from Butterworth to Hermoza & Co. in Mazatlán, he complained that some of his surplus mercury that was sold to parties in London (probably the Rothschilds) was finding its way back to Mexico from Vera Cruz, fur-

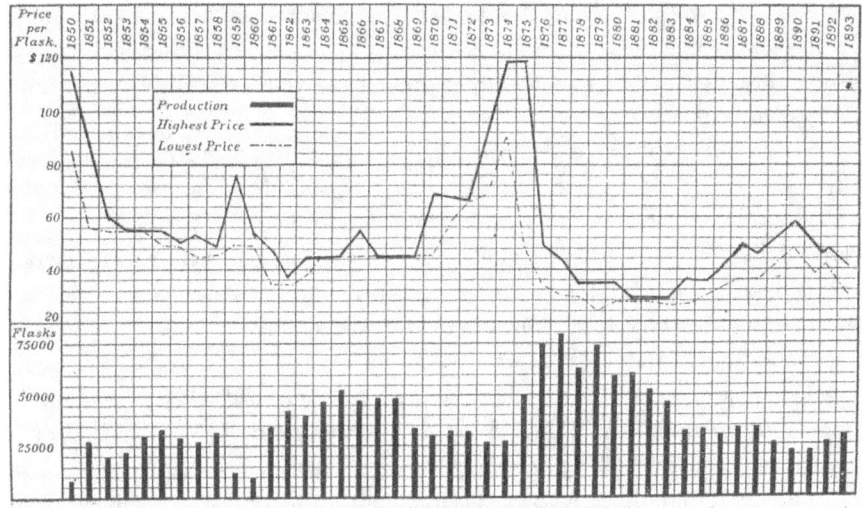

FIGURE 2.2 Production and Price of Quicksilver in the United States

Major events in the history of the mercury industry in California are visible in the chart. During the first two years of New Almaden production (1850–52), the price of mercury dropped by half. The 1859–60 injunction closing the New Almaden Mine curtailed production and resulted in a spike in price. The Magic Quicksilver Ring, an industry trust, gradually suppressed production in the late 1860s and early 1870s, and that, coupled with a rise in demand, caused the price to reach extraordinary peaks in 1873 to 1875. After the price boom came a production boom and price bust as both the original mines increased production and new mines developed. (Day, *Mineral Resources of the United States* [1893].)

ther depressing his sales. He advised Hermoza & Co. to work off its large stock even if a reduction in price was necessary, to counter both the Spanish and the parties in London and secure the market of western Mexico for New Almaden.[31]

The predicament of the QMC during these years was not entirely of its own doing. Both the Rothschilds and Barron & Co. hoped to secure deals for marketing the New Almaden mercury and even to control the mine, and each made strategic moves to achieve these goals.[32] The London Rothschilds, through their New York agent August Belmont, made multiple attempts at securing a deal with the QMC. For a few years in the mid-1860s the Rothschilds did not have a deal with the Spanish government, which instead was selling its own quicksilver. In a letter from Belmont to the QMC dated June 2, 1865, he wrote, "If the Quicksilver Mining Co. can afford to sell at that rate (7 pounds or less per flask) I think that we could soon monopolize the markets of the world and bring the Spanish government which now works the mine for its own account to stop operation altogether."[33] In July 1865 the Rothschilds proposed a deal in which they would buy 25,000 flasks a year if the QMC would limit production to that amount.[34]

The deal that the QMC ended up making, however, was with Barron & Co. and began on April 1, 1866. Under this deal the QMC sold them the total production of the mine, up to 50,000 flasks, delivered prior to April 1, 1868. Barron & Co. paid thirty dollars in gold per flask, and made additional arrangements whereby the payments to Barron, Forbes & Co. for the sale of the mine in 1863 could be made in quicksilver to Barron & Co.[35] One can almost read a sigh of relief in Butterworth's copybook from April 1866 as he wrote to his former agents around the world advising them that he was no longer selling quicksilver, that he had sold the total output of the mine to one party.[36] This deal was an admission by the QMC that it was not capable of successfully selling its own quicksilver. After 1866 the QMC was only in the business of running a mine, and this the company did for many decades to come, while missing out on the extraordinary wealth that would be made by controlling California quicksilver and offering this control as leverage in forging business relationships in California.

The 1866 QMC Annual Report detailed the relationship it had established with Barron & Co. The most telling sentence of this deal reads, "This contract was made with Mr. W. E. Barron of San Francisco, and the due performance thereof on his part was guaranteed by Mr. D. O. Mills and Mr. Wm. C. Ralston."[37] Mills and Ralston, major figures in the history of California and the West, had just founded the Bank of California on July 1, 1864, and this bank served as the catalyst for their varied business enterprises. Thomas Bell and William Barron were founding partners in the bank, and thus part of the "Bank Crowd," a group of influential Californians who came together through the bank to form and develop mostly very profitable business interests, some of the most profitable being the control of the Comstock Mines and bullion mills, and the railroad connecting these mines to California.[38] Although there have been many histories written concerning Ralston, Mills, the Bank Crowd, and the Bank of California, there has been little if anything written on the importance of William Barron and Thomas Bell, and their control of mercury, to the Bank Crowd.[39]

Late in his life, in an interview for H. H. Bancroft, Thomas Bell said the following about his dealings with Ralston and the Bank Crowd in the mid-1860s:

> The agency of the Bank (the Bank of California) at Virginia City having made advances in 1866 to a number of quartz mills there; when the first Bonanza in the mines was worked out, and ore scarce, the mills became bankrupt, and the Bank took

possession of them; these mills were bought from the Bank by Sharon, Mills, Ralston, Thomas Sunderland and Thomas Bell, who established the Union Mine and Milling Company; this in subsequent years became the nucleus of enormous profits in connection with the Virginia & Truckee Railroad, in which the same parties were owners.[40]

Quicksilver was a major basis of the initial relationship between Barron and Bell, and the other members of the Bank Crowd. Barron and Bell brought to this relationship the quicksilver that was crucial to the operation of the mills. Together, then, the Bank Crowd owned the mills and controlled the quicksilver supply that was necessary to run them.

As boosters of San Francisco and California, Barron & Co., Ralston, and the rest of the Bank Crowd worked together to develop the state and in the process kept their money on the West Coast. By contrast, the QMC, as an East Coast company, siphoned any profits from the West.[41] Due to the QMC's financial troubles, however, it had to borrow large sums from the Bank of California, and its loans put it under the control of banking powers. Before 1866 the QMC also borrowed money from the London Rothschilds, to be repaid in shipments of quicksilver to London. After 1866, the deal with Barron & Co. relieved some of their debt worries, particularly the payments that were due to Barron, Forbes & Co. for the purchase of the mine. This transfer of their indebtedness to Barron & Co., together with new loans from the Bank of California, led to domination of the QMC by the Bank Crowd. On December 2, 1867, Ralston wrote to Butterworth with tough terms:

> The balance due this Bank from your Company is now $435,000. For a portion of this amount, say $200,000, our Mr. Mills, at this time in New York, has made a demand upon your Board of Directors, to which he finds them unable to respond... In view of these circumstances and of the fact that a large portion of our claim is uncovered, we have to request that you will, with as little delay as possible, proceed to reduce the surplus ore now on hand at the mine, and place in our possession the product in Quicksilver; to be realized as interest may appear.[42]

The loan allowed Ralston to put pressure on the QMC to provide him and his Bank Crowd partners with low-cost quicksilver, strengthening their control of both the California and Pacific markets.[43] When Butterworth and the QMC directors realized in 1866 that they did not have the ability to sell their own mercury in the global "retail" markets, they resigned their grand goal of a quicksilver monopoly and instead made deals simply to survive. The model that the QMC touted in its Annual Reports—of a mine that controlled the bullion markets of the world—was possible when Barron, Forbes & Co. owned New Almaden because it could produce, sell, and use mercury. Barron, Bell and the Bank Crowd demonstrated that the ownership of the New Almaden Mine did not matter as much as the control of the mercury once it was produced.

The 1860s: The Magic Quicksilver Ring

On March 31, 1868, the first contract between the QMC and Barron & Co. for the sale of mercury ended. Barron and Bell declined to enter a new contract for direct sales with the QMC because of a new force in the industry: the Redington Quicksilver Mine in Napa County, first developed in 1862 and reorganized and developed by San Francisco capitalists in 1867 to greatly boost production. Instead Barron and Bell formed a combination of the three major producers in California: the New Almaden, the New Idria, and the Redington. This combination, secretive at the time but suspected and exposed in the press, was generally referred to as the Magic Quicksilver Ring.

Barron & Co. was part owner of New Idria with other members of the Bank Crowd.[44] These parties also had solid control of New Almaden through financial manipulation, so fear of the Redington Mine was the primary reason for this new combination.[45] By the late 1860s the Redington Mine had emerged as a potential spoiler to the plans of Barron, Bell, and the Bank Crowd. The Redington was owned and operated by a group of men independent of the Bank Crowd who had strong business relationships with the gold and silver mines of California and Nevada. Unlike the QMC, which could never establish the necessary trading relationships for the sale of mercury, the Redington Mine was named after a major investor—John Redington—who had come to California in 1848 and who was the owner of the prominent wholesale drug firm of Redington & Co., with connections and outlets throughout California and the West as well as a branch house in New York.[46] Although the owners of the Redington Mine had no facility for international sales, they did have a well-established network for mercury sales in mining regions in the West and for sales in New York. Barron and Bell worked hard to negotiate a combination with the Redington owners and eventually succeeded in making a deal in which each major mine had an annual quota, with Barron & Co. being the sales agent for all exported mercury, while the Redington owners were credited with all sales for California and Nevada (although Barron & Co. was the agent for these sales). The combination limited the New Almaden Mine to 24,000 flasks annually and New Idria and Redington to 10,000 flasks each.[47] This deal allowed Barron & Co. to limit production in the state in order to protect the price on world markets.[48] It also allowed the company to bring new mines, especially the Redington Mine, into its fold and avoid outright competition.

The three mines and Barron & Co. signed on to this combination for a period of two years. An additional provision of the contract made purchasing agreements with all the minor quicksilver producers, effectively eliminating the possibility of establishing any new mines in the state. The combination agreement positioned all California producers as one trading entity on the global quicksilver market, equal to the Almadén Mine in Spain, and made Barron & Co. equal to the Rothschilds in world quicksilver sales. In 1869 the U.S. commissioner of mining statistics, Rossiter W. Raymond, wrote in a government report:

> The quicksilver trade of the world is substantially an armed truce between Spain and California. The mine of Old Almadén, in Spain, supplies the market of London and a large part of Europe, and ships its product as far west as the city of Mexico. Until recently, it also controlled the great Chinese market, but Mr. Butterworth, shipping 10,000 flasks to Hong Kong, and selling at far below the cost, forced the re-shipment to Spain of all the Spanish quicksilver; and the market has since been in his hands. The same tactics on the part of Spain keep him from the London market; and the two great producers are thus forced to divide the world between them.[49]

This combination engineered by Barron & Co. was a new type of organization in the history of quicksilver. Whereas the idea of monopoly control was consistent with the long history of mercury, the need to control a whole region—in this case California—with many mines and prospective mines was something new. Although the Spanish Crown had used its royal power to forbid the development of additional quicksilver mines, the California speculators had to use markets (and tools such as monopolies to control markets) to exercise the power to limit quicksilver mining. Their tool of control was money, although the art to their quicksilver monopoly was in achieving and maintaining a complicated balance of supply, demand, price, and power. Their goal was to reap as much reward as possible from their monopoly while still holding the monopoly together.

The first two-year combination (April 1, 1868, to April 1, 1870) was a learning experience for all the parties to the contract. Barron & Co.—who had arranged to buy mercury at a set price—realized advancing prices and decreasing supply, the perfect combination for profit. For its members, the combination worked well as a tool they could use to control the changing California industry. For the New Almaden Mine the combination had been a life-support mechanism, allowing survival of its troubled financial situation by guaranteeing an income. The mine was not able to prosper, however; Barron & Co. and the Bank of California arranged it so that the QMC prospered only enough to continue producing the flasks of mercury. For the Redington Mine the combination produced wealth but not the great wealth it knew could be made and that it knew Barron & Co. was making with its quicksilver. As the odd man out in the combination, the Redington Company knew that the combination had been formed in large part to control it, and its owner-managers looked to improve their situation in the future. The New Idria Mine, under the direct control of Barron & Co., was manipulated by Barron & Co. as the company desired. As it happened, the New Almaden and Redington Mines produced significantly less than the quota amounts dictated by the combination, whereas the New Idria Mine slightly exceeded its quota.[50]

The combination contract was significantly revised when it was renewed on April 1, 1870. Both the New Almaden and Redington Mines had their situations modified, but from very different positions: New Almaden from desperation and the Redington Mine from strength. New Almaden had two problems, starting with a financial crisis in 1870, largely from its indebtedness to the Bank of California. Butterworth had made a deal with D. O. Mills, the president of the bank, at the time of the first combination by which the QMC sold all its mercury in foreign markets (reputedly 20,000 flasks) to Mills for thirty-two dollars a flask, in addition to the 2,000-flask-per-month sales made as QMC's quota within the combination.[51] For the QMC this deal meant a reduction of its debt to the Bank of California, while a loss in potential for any profit from this mercury. The second problem was the decreasing richness of the ore coming from the mine.

The Redington Mine owners knew that the New Almaden Mine and the New Idria Mine were controlled by the same parties. They understood the benefits of the combination, but wanted to improve their position in it. They eventually negotiated a ten-year contract whereby they would sell all the flasks they produced, up to 10,000 annually, to Barron & Co. and D. O. Mills for a set price, reputed to be about forty dollars each (Figure 2.3).[52] Unfortunately for the Redington Mine, the market price for quicksilver rose to nearly seventy dollars a flask in 1872 due to decreasing production from the New Almaden Mine, greater worldwide demand, and the higher price charged by the Rothschilds for quicksilver from Almadén. As Rossiter Raymond wrote in *Statistics of Mines and Mining in the States and Territories West of the Rocky Mountains for the Year 1870*:

> The owners of the Redington Mine, who, under their contract, still hold exclusive control of the local sale, now have the satisfaction of delivering the quicksilver to Barron and Mills at $40 per flask, and reselling it under their orders for $68.86. Strangely enough, the product of their mine, which started off at the full limit, has, as the price increased, gradually dwindled.[53]

The other major change at the time of the second combination was that Samuel Butterworth left the Quicksilver Mining Company to be the president of the North

Total product of quicksilver in the United States.

[Flasks of 76½ pounds, net.]

Years.	New Almaden.	New Idria.	Redington.	Sulphur Bank.	Great Western.	Napa Consolidated. (a)	Great Eastern.	Mirabel. (b)	Various mines.	Total yearly production of California mines.
1850	7,723									7,723
1851	27,779									27,779
1852	15,901								4,099	20,000
1853	22,284									22,284
1854	30,004									30,004
1855	29,142	Production from 1858 to 1866—no yearly details obtainable—included in production of various mines. 17,455 flasks							3,858	33,000
1856	27,138								2,862	30,000
1857	28,204									28,204
1858	25,761								5,239	31,000
1859	1,294								11,706	18,000
1860	7,061								2,939	10,000
1861	34,429								571	35,000
1862	39,671		444						1,885	42,051
1863	32,803		852						6,876	40,589
1864	42,489		1,914						3,086	47,489
1865	47,194		3,545						2,261	53,000
1866	35,150	6,525	2,254						2,621	46,550
1867	24,461	11,493	7,862						3,184	47,000
1868	25,628	12,180	8,685						1,234	47,728
1869	16,898	10,815	5,018						1,580	33,811
1870	14,423	9,888	4,546						1,220	30,077
1871	18,568	9,180	2,128						2,810	31,686
1872	18,574	8,171	3,046						1,830	31,621
1873	11,042	7,735	3,294		340				5,231	27,642
1874	9,064	6,911	6,678	573	1,122				3,388	27,756
1875	13,642	8,432	7,513	5,372	3,384		412		11,489	50,250
1876	20,549	7,273	9,183	8,367	4,322	573	387		22,063	72,716
1877	22,906	6,816	9,399	10,993	5,856	2,229	505		20,101	79,395
1878	15,352	5,138	6,686	9,465	4,963	3,049	1,366		17,361	63,880
1879	20,514	4,425	4,516	9,249	5,333	3,605	1,455		23,587	73,684
1880	23,465	3,209	2,139	10,706	6,442	4,416	1,279		8,270	59,926
1881	26,060	2,775	2,194	11,152	6,241	5,552	1,005		5,812	60,851
1882	28,076	1,953	2,171	5,014	5,179	6,842	2,124		1,379	52,732
1883	29,000	1,606	1,894	2,612	3,869	5,890	1,069		185	46,725
1884	20,000	1,025	981	890	3,292	4,307	332		1,186	31,913
1885	21,400	1,144	385	1,296	3,469	3,508	416		427	32,073
1886	18,000	1,406	409	1,449	1,949	5,247	735		786	29,981
1887	20,000	1,890	673	1,490	1,446	5,574	689	1,543	520	33,825
1888	18,000	1,320	126	2,164	625	5,024	1,151	3,848	992	33,250
1889	13,100	980	812	2,283	556	4,590	1,345	1,874	924	26,464
1890	12,000	977	505	1,608	1,334	3,429	1,046	1,290	737	22,926
1891	8,200	792	442	1,375	1,844	4,454	1,660	1,686	2,451	22,904
1892	5,563	843	728	1,393	5,867	7,272	1,630	3,208	1,484	27,993
Total	930,122	132,906	100,923	87,451	68,433	75,559	19,270	13,449	188,214	1,616,353

a Includes Ætna.
b This mine previous to 1892 was called Bradford.

FIGURE 2.3 Production History by Mine of Quicksilver in California

In the typology of mercury mines presented in chapter 3, Figure 3.4, New Almaden and New Idria are the original mines, the Redington is a type by itself, and the remainder of the mines are new mines. (Day, *Mineral Resources of the United States* [1893].)

FIGURE 2.4 Malakoff Diggings

This photograph, by Carleton Watkins, shows hydraulic miners for the North Bloomfield Mining Company, a company partially owned by Thomas Bell, at work washing down alluvial deposits bearing gold. The resulting slurry was then run through mercury amalgamation mills. (Courtesy of the Hearst Mining Collection of Views by C. E. Watkins.)

Bloomfield Gravel and Mining Company, a major hydraulic gold mining company in the Sierra foothills.[54] In 1866 Ralston, with a group of investors including Bell and Butterworth, had consolidated a number of small hydraulic mines into the North Bloomfield Mine.[55] The company operated the Malakoff Diggings, the largest hydraulic mining area in the state (see Figure 2.4). Although Butterworth had been outmaneuvered at the quicksilver game, he was enough of a businessman and pragmatist to take new opportunities and join with Ralston and the Bank Crowd in an enterprise where his influence and connections with quicksilver allowed him to profit from the metal in a more valuable way.[56]

By the spring of 1870 Butterworth had put his nephew, James B. Randol, into his old position as manager of the New Almaden Mine. Randol had formerly been the secretary of the QMC in New York, and in Randol, Butterworth had a reliable and capable manager. However, he was not a powerful figure outside of the mine. Randol's instructions were to make the mine self-sustaining, and the mine records show that he focused on this objective. All the sales of quicksilver were secured— Randol had little to do with sales and the price of quicksilver. Whereas Butterworth's correspondence during his five years as manager is wide-ranging, international, and full of large-scale wheeling and dealing, Randol's correspondence is locally focused and detail oriented. He took control of all mine expenses, and was deeply involved in comparison shopping for mine supplies such as candles and blasting powder. Randol also fired many existing mine supervisors in an effort to bring his management principles to the operation of the mine.

The Magic Quicksilver Ring managed to control mercury in California from 1868 until April 1, 1873; during that time, world quicksilver supplies decreased and fell more in line with demand, and the price on world markets increased. These trends, induced by the combination, were augmented by an increase in demand both domestically and in foreign markets. This combination was immensely successful in benefiting the Bank Crowd at the expense of nearly all the other parties in the mercury trade, including two of the three quicksilver mines and the industries that needed to purchase high-priced mercury on the domestic and foreign markets. The exception was the New Idria Mine. Barron and Bell had direct control of the production totals at New Idria, and although it was a rich mine capable of producing large quantities of quicksilver, Barron and Bell regulated production at the mine to suit the combination's business in trading mercury. In the combination, New Idria had a higher quota than was seemingly warranted by past production, and as prices rose mine production was increased and the mine was very lucrative for Barron, Bell, and the Bank Crowd in the combination years.

The 1870s: The Quicksilver Boom

When Butterworth and the QMC decided to quit selling their own quicksilver in 1866 and instead contracted with Barron & Co., it marked the beginning of an era in which quicksilver became an integral part of the vertical monopolies constructed by William Ralston and the Bank Crowd.[57] The Comstock experienced early bonanzas in 1866, 1867, and 1868, and Barron and Bell capitalized on sales of mercury during these years.[58] Then, between 1869 and 1871 there was a major decline on the Comstock, and the Bank Crowd bought up distressed amalgamation mills in the hopes of another boom. The Bank Crowd's gamble paid off in an all-out bonanza of new ore in 1872–73. Bell (Barron had died in 1871), Ralston, and the rest of the Bank Crowd enjoyed the tremendous profits to be had from controlling the bullion production from one of the major bonanzas in the history of the West, in part because they controlled the supply of mercury, which they could liberally sell to themselves while charging any unaffiliated mill high prices, if they made mercury available to the mill at all.

Mercury's place in the vertical monopoly of the Bank Crowd persisted through the three subsequent combinations, the last ending on April 1, 1873.[59] After this date the containment of the industry that had been enforced by the combinations failed, and the industry—which had been experiencing rising prices—was launched into a price boom that lasted into 1875 (see Figure 2.2). When the mercury combination failed, the markets—as the tool of control—also broke down and took a decade or more to regain equilibrium. Helping to set the conditions for the price boom and the combination's failure was the low production of mercury from the California mines from 1869 to 1874. This decrease was due in large part to the reduced production from the New Almaden Mine. Jimmie Schneider, in his history of the New Almaden Mine, attributes this decrease to legitimate mining difficulties that included low-grade ore and the inability to process this ore.[60] Up to about 1868 the New Almaden Mine produced very rich ore, and the reduction plant—the furnaces and condensers—was tailored to run on this ore. However, when the rich ore ran out and the mine was left with large quantities of lower-grade ore, the production plant technology had to be modified. The solution to reducing large quantities of low-grade ore, which was in place by 1875–76, was the fine-ore continuous furnace. The development of this technology, however, did not come soon enough and in the early 1870s with the price booming, the managers of New Almaden were desperate to produce more mercury. In a letter dated December 16, 1873, months after the end of the last combination, QMC president Daniel Drew wrote to mine manager Randol highlighting how they had been taken by Barron and Bell:

> I notice that the New Idria Mine heretofore regarded as compared with us of minor importance, has delivered for the past six months nearly as much metal as we, and for the past three months were 600 flasks in excess. The year prior we produced and sold to the principal owner of the New Idria over 18,500 flasks at a low figure, which resulted in a very handsome profit to them, while their production was comparatively low—now that the price is remunerative it is particularly unfortunate for us that our production should be so very light.[61]

When the Magic Quicksilver Ring reached a breaking point in 1873, so did public opinion concerning the industry. The expense of mercury and the tightness of the supply were blamed in the press as a partial cause of the ongoing economic downturn. Even after the last combination expired on April 1, 1873, the people behind the combinations continued to reap financial rewards from quicksilver. Profits, however, were based on the relative scarcity of mercury due to low production, and the fact that the mercury that was being produced was mostly made available within the vertical monopolies of the mining West. Domestic sales, mainly to the Comstock mills, took precedence over foreign markets. From 1855 through 1864 nearly 80 percent of all California mercury was exported. In 1868, during the first Comstock Boom, nearly 68 percent was exported, whereas in 1873 and 1874 exports accounted for only about 23 percent of domestic production.[62] Exposé newspaper articles detailed the inner workings of the quicksilver industry and the depressive effects the limited and controlled supply of mercury was having on the mining industry; the articles called for the government to remedy the situation. Certain mills, including the Carson Mill Co.—which was not owned or in partnership with the Bank Crowd—went bankrupt due in large part to the high cost and unavailability of quicksilver.[63] Such failures were prominently featured in the press, and this coverage pressured elected officials to address the problem. In August 1873, California governor Booth gave a speech attacking the quicksilver monopoly:

> For years quicksilver has been a monopoly which, transcending the boundaries of geography, has had the whole world for the field of operations. A combination was formed some years ago, and I believe still exists, among the great quicksilver companies, with the Rothschilds at the head, by which the production of each was limited to a fixed amount, and the particular division of the globe which each might supply was duly assigned. These potentates divided the earth into commercial kingdoms, and enthroned themselves as kings; and now, let any man endeavor to develop a quicksilver mine in this State, he will find himself harassed—perhaps ruined—by causeless litigation instigated by the monopoly.[64]

The easing of tariffs on the importation of quicksilver was one method that government officials explored in reducing the quicksilver crisis in the mining industry. In 1873 the tariff was reduced from 15 percent to 5 percent, and on February 8, 1875, the United States lifted all tariffs on the importation of quicksilver.[65] As well as manipulating the tariff, there was also talk of a bill in the federal Congress to regulate the price of quicksilver. Bell addressed this bill in a letter to New Almaden manager J. B. Randol on February 13, 1874:[66] "If such a bill became law and the courts could enforce it—there would be no other course for the owners of quicksilver mines but to shut up shop—and then the dear public would have to go to London for the article—it's too absurd."[67]

Emerging from the price boom of 1873–75, and the public battles to solve the mercury crisis, was the fact that many new quicksilver prospects were being discovered throughout the state of California. Although few of these mines produced significant quantities of quicksilver until 1876, their presence worried Thomas Bell and provided hope for newspaper commentators looking for an end to the crisis. In 1874 Thomas Bell commissioned a study of the new quicksilver prospects, many of which were in Napa and Sonoma Counties.[68] This report predicted that there would be 2,500 to 3,000 flasks per month from these mines in the near future, and to Bell—who knew the market for California quicksilver to be not more than 5,000 flasks per month—these new mines and their production potential were ominous.

These new mines were the direct result of Bell's and the Bank Crowd's loss of control over the quicksilver industry. The price boom, for the first time in the history of the California mercury industry, made mercury profitable from direct sales. A person with a cinnabar outcrop could dig out some ore, refine it in a simple furnace, and have hope of selling the resulting mercury at a profit. The boom in new quicksilver mines soon became a popular item in the press, and in quicksilver regions of California (up and down the Coast Ranges) the quicksilver mine boom fueled the development of existing towns and a number of quicksilver boomtowns.

The quicksilver mining boom was different from other mining booms in the state because the deposits were in the Coast Ranges, not far from developing towns and near San Francisco and the Bay. Although many of the new prospects were in rugged terrain in the hills and mountains surrounding Napa and Sonoma Valleys and Clear Lake, few of the prospects were more than a few miles from the developing towns of these counties. For these already blessed counties, quicksilver was an additional bounty. Many of the people who located new mercury prospects were not prospectors by trade. The *Mining and Scientific Press* printed in 1875: "The quicksilver excitement as well as its price may be said to have kept up to the highest pitch during the whole year... Honest farmers have dropped the plow and taken to the pick, and even the female part of the population in some places, have obtained prospectors' outfits and scoured the hills in search of the precious metal."[69] The people who discovered cinnabar outcrops probably imagined that they now had a way to tap into a mining bonanza. Small-town newspapers, in the business of attracting new settlers and new development, loudly proclaimed the quicksilver riches that surrounded their towns and made wild projections of the riches these mines would produce.

Although most of these new prospects were not developed until after the price boom ended in 1875–76, the older mines all achieved significant financial gain during the boom. For the New Almaden Mine, the high prices of the 1870s brought a level of prosperity despite low production totals. The Redington Mine was a winner during the quicksilver boom, making up, in part, for its disastrous contract from 1870 by hitting a production peak while the price of mercury was still high.[70] New

Idria, not surprisingly, was the most lucrative of the California mercury mines, its production having been manipulated by Thomas Bell to peak during the boom years.

The 1870s: The Quicksilver Bust

By 1875 the boom was over. Prices were on the way down as the Comstock Boom ended and as mercury production increased. But the quicksilver boom had lasting effects, changing the nature of the mercury industry in California. Before the boom, Bell and the Bank Crowd had achieved monopoly control over the industry in California by controlling the product from the three major mercury mines. During the boom the three major mines together produced much less mercury, receding in importance, while many new mines emerged. A few of these new mines proved to be significant, long-term producers, including Sulphur Bank, Great Western, Great Eastern, and Napa Consolidated, all north of San Francisco Bay. Most of the new mines, however, were very small and were profitable only during the price boom. As a group the larger new mines about equaled New Almaden's production in the late 1870s. Whereas New Almaden had averaged well over 80 percent of California mercury production from 1861 to 1865, and over 50 percent from 1868 to 1873, it averaged less than 30 percent of California production from 1874 to 1879. After 1875 there were dozens of mine operators and investors, all with quicksilver prospects and all trying to sell their product. During these years the mercury market was as open as it would ever be in nineteenth-century California.

The mercury price bust was directly tied to the Comstock market crash.[71] Wild stock speculations, particularly on Comstock Mine stocks, on the San Francisco Mining Exchange led to the crash in January. The crash was due in part to William Ralston's attempt to gain control of the Ophir Mine on the Comstock, an ultimately unproductive plan that drove stock prices to unprecedented heights. For years before, however, Ralston had been overextending himself in speculative schemes of all kinds—the crash of the Comstock stocks was one more large loss on top of others. On August 26, 1875, the Bank of California was forced to close its doors for lack of funds, and when an inquiry was started the next day, the extent to which Ralston had been appropriating bank funds for his own use became clear. Ralston was asked to resign from the bank and that same day drowned in San Francisco Bay. D. O. Mills, William Sharon, and Thomas Bell provided guaranties to the bank's major creditors and managed to reopen the bank.[72]

By March 12, 1875, Bell had written to Randol that the quicksilver market was going from bad to worse in large part, according to Bell, because the new mines were making low offers without effecting sales.[73] By March 22, the price had dropped so much that Bell told the QMC that he was not going to accept any more quicksilver unless the price, set before the crash, was renegotiated.[74] Negotiating and renegotiating quicksilver deals devoured Bell's attention for the next few years. He now had five major independent quicksilver producers to deal with, rather than the one (Redington) he had prior to the collapse of the combination. In 1876 Bell's letters show that he was constantly attempting to negotiate new combinations among the major quicksilver producers. The new Comstock kings—Flood, Fair, and O'Brien—who rose to prominence as the influence of the Bank Crowd receded, picked up whatever they could of Ralston's shattered empire and in the process became the major quicksilver purchasers on the Pacific Coast. The Redington Mine had a long-term lucrative deal to sell them mercury, and other new mines all fought over each other for their business. Flood and O'Brien took the initiative, however, and in December 1876 negotiated to purchase a half interest in the Guadalupe Mine, near the New Almaden Mine.[75] Unfortunately for Bell these combinations either never materialized or collapsed soon after being established, because one or more of the parties broke the contract. In addition, a small mine would from time to time suddenly hit rich ore and demand to be included in the combinations, throwing previously negotiated arrangements into disarray. In a letter signed by Bell on October 10, 1876, he complained that two of his new associates (and adversaries), Anthony Zellerbach and Tiburcio Parrott, were not willing to come into a combination on equal terms with the other parties.[76]

Bell faced a serious challenge to his preeminence as the Quicksilver King in January, 1877, when the Merchant's Exchange Bank (MEB) of San Francisco made an attempt to organize a quicksilver board to sell all California quicksilver.[77] As agents for a number of the new mines, the MEB argued that the present plan was not working and the sellers of quicksilver should themselves take control of the California mercury market. The MEB proposed that a board of three directors (on three-month terms) be appointed to set the price and determine the rules. Sales would be made in the order that the quicksilver was received. It proposed that Bell could be one of the rotating directors.[78] Bell was incensed at this proposal. He wrote to J. B. Randol on January 24, 1877: "The fact of the matter is—*there is a rebellion*. They see I have more facilities for making sales than they and they want me to divide with them. If the arrangement is broken up—and I have no doubt it will be—we shall see very low prices."[79] Bell faced a challenge to the model of quicksilver sales that he had always known and that he had carefully constructed for his own benefit. Bell wanted to preserve his status as the Quicksilver King, the rightful heir of Barron, Forbes & Co., whereas the MEB wanted to set up a rotating board of sellers to manage the industry. The MEB proposal died quickly; Bell's control of New Idria and his dominant position with New Almaden made him too powerful. It was not until after Bell's death that a quicksilver marketing board organized much the way the MEB proposed was established.[80]

By the end of 1877, Bell succeeded in forming a combination. His major task was to limit quicksilver production, and his letters are full of attempts to get the various mines to agree on amounts. In powerful figures such as Zellerbach (who had quicksilver mines because he needed a supply of mercury for his paper mills) and Parrott (who had a major trading house and his own bullion interests), Bell was dealing with people who were not only in the business of producing quicksilver, but who had their own vertically integrated empires of which quicksilver was often only a modest part. These people were interested in actually using quicksilver as a tool to wealth as much as if not more than attempting to control the price on world markets.[81] These contentious dealings lasted through the 1870s, as the mines continued to fight among themselves and with Bell. However, by the end of the decade, with prices very low and production totals very high, there was a weeding-out within the industry. Many of the smaller mines, often developed with little financial backing, either closed because they were not profitable or were forced to close by the new combinations that could undersell them until they collapsed.

The 1880s: The Decline of the Industry

The mercury industry suffered three major blows in the early 1880s. The first came in 1880 when most of the new mines that had been developed during the boom years were closed because they employed Chinese miners, against a provision of the new state constitution. The Napa Consolidated Mine was the most prominent mine to close, being the third-largest quicksilver mine in the state, and employing mostly Chinese workers. However, Tiburcio Parrott, one of the major players in the industry and the owner of the Sulphur Bank Mine, challenged this provision of the constitution in court, and eventually won.[82] Some of the more major mines reopened after the Parrott decision, including the Napa Consolidated and the Great Western, and the quicksilver industry retained Chinese miners statewide. However, this case highlighted the fact that quicksilver mining, unlike gold and silver mining, employed Chinese miners, and the industry was a target for anti-Chinese sentiment throughout the 1880s.

The quicksilver industry took a more financially devastating hit on January 7, 1884, when Judge Sawyer of the Ninth District Court issued a permanent injunction against the North Bloomfield hydraulic gold mine. For years farmers and townspeople downstream from the hydraulic mines had been organizing to stop the mines.

The direct result of this injunction was that one of the major domestic consumers of mercury was out of business, and other hydraulic mines in the Sierras were operating on borrowed time. The hydraulic industry, however, moved elsewhere in the West, and hydraulic mining continued to be a significant purchaser of mercury. For Thomas Bell, one of the founders of the North Bloomfield Mine, the court-issued injunction was a major blow.[83] Although it is true that the way to wealth with mercury was to be a double winner by controlling both the mercury supply and use, it is also true that when either mercury production or use was lost, then one could also be a double loser. Bell lost both the income from a major gold mine and the income from the sale of mercury to the mine. Bell, however, died a wealthy man,[84] and even after the setbacks in quicksilver of the 1880s he continued to search for opportunities to create capital, as he had throughout his life.[85]

The final blow to the industry came in the late 1880s, when a cyanide process replaced mercury amalgamation as the primary means of refining gold and silver.[86] By around 1890 most gold and silver refining was by cyanide, and the domestic market for mercury for amalgamation mostly disappeared. Although a blow to the quicksilver industry, the loss of this domestic market was not devastating to sales because most of the quicksilver was still sold internationally, and many of these sales were not related to gold and silver processing. What was devastating to the mercury industry was the loss of its influence in the gold and silver mining industries. Mercury was no longer a tool for achieving great wealth—it was instead reduced to a simple commodity produced for sale, worth its trade value on international markets. No longer would wealthy power players battle for control of the industry.

The successful manipulation of mercury production, trade, and use as a tool to power and wealth, passed down from the Spanish and British and reshaped for California, guided the development of the quicksilver industry in California until the development of cyanidation. The California mercury experience was particular, however, in that for the first time mercury was controlled by capitalists who were involved with mercury for financial gain and the power that came with having money. In the American West capitalism was the defining characteristic of the American empire, and the historical role mercury had played in empire was ingeniously adapted to the rapidly changing forms of capitalism in the American West.

Since the sixteenth century, mercury had been a political tool of Crown powers, used to bolster the power of the Crown. Although the Rothschilds had reestablished the mercury trade following the collapse of the Spanish Empire, their business in mercury was still closely tied to the British Crown and the expansion of the British Empire. Barron, Forbes & Co. further separated the mercury trade from Crown powers with the New Almaden Mine on the frontier in California, and by developing and then controlling the Pacific trade in mercury. However, it was the new business model developed by William Barron, Thomas Bell, and their partners in the Bank Crowd—of controlling mercury trade and leveraging the power that came from its use in their vertical monopolies—that made mercury a tool of capitalists, and helped make the members of the Bank Crowd wealthy.

Notes

1. In 1851, the year New Almaden doubled world quicksilver production, Alexander Forbes retired to England, ending his day-to-day operation of Barron, Forbes & Co. See the introduction in Forbes, *California: A History of Upper and Lower California*, xxi. At this time Alexander Forbes was in his seventies. Eustace Barron bought out many of his assets in the company, although Forbes retained an interest in the New Almaden Mine. For his part Barron began to enjoy his great wealth. In the early years of operating the New Almaden Mine he bought Santa Cruz Island, one of the Channel Islands off the coast of Santa Barbara. Eustace Barron bought Santa Cruz Island from Andrés Castillero, the same man who had originally claimed the New Almaden Mine. The island was both a private retreat, stocked with sheep he brought from Great Britain, and perhaps a safe port for company trading vessels where they could dock and load and unload out of view of any nation's officials, yet be in easy reach of the major Pacific ports of California and Mexico.

2. In the mid-1840s, James Bolton had been acting consul for John Parrott, who would become one of the richest men in California through varied business concerns, most important a trading house—Parrott & Company—and through speculation in San Francisco real estate. About 1850 Parrott became part owner of the New Almaden Mine, presumably because he was well respected and powerful, and aided in the operation and development of the mine.

3. Eustace Barron and Alexander Forbes founded Barron, Forbes & Co. about 1823 in Mexico. This firm was the majority owner of the New Almaden Mine until 1863. James Bolton and William E. Barron (a son or nephew of Eustace) founded Bolton, Barron & Co. about 1850 in San Francisco to handle the sale of mercury from New Almaden. While closely affiliated with Barron, Forbes & Co., this firm was a separate entity. Thomas Bell was a member of this firm, but he was not a partner. In 1858, James Bolton left the firm (or was forced out due to his scandalous dealings with the lands of Mission Dolores), and Thomas Bell became a partner with William E. Barron in the firm of Barron & Co. William E. Barron and Bell, with members of the Bank Crowd, bought control of the New Idria Mine in 1858. When the New Almaden Mine was sold in 1863, Barron & Co. lost its right to New Almaden mercury sales, but regained these sales through negotiation with the new owners under Samuel Butterworth. William E. Barron died in 1871, and following his death Thomas Bell renamed the firm Thomas Bell & Company, with George Staack as agent and later a partner in the firm. Upon Bell's death in 1892, Staack became the managing secretary of the Eureka Company, an

industry trust built upon the remnants of Bell's firm, set up to sell mercury for the major mines of the industry. A notice appeared in the *New York Times* on October 17, 1893, announcing the liquidation of Barron, Forbes & Co. after nearly seven decades of operation. The Eureka Co., at least officially, disbanded in 1914 following the passage of national antitrust legislation. For more on Barron, Forbes & Co., see Mayo, *Commerce and Contraband on Mexico's West Coast in the Era of Barron, Forbes & Co., 1821–1859*.

4. For more on the importance of reputation in banking and merchant houses in early California, see Lynne Pierson Doti and Larry Schweikart, *Banking in the American West*.

5. J. B. Randol, *Papers Relating to Quicksilver Mining, Ca. 1849–1894*, vol. 2, folder 1. This early information on Bolton, Barron & Co. comes from a transcription written at New Almaden on March 14, 1876. The reason for this transcription is unknown. This document begins, "I have before me a closely written book of 355 pages, size 8 inches by 13¼ inches, on the back it is marked Dividends and Account Sales No. 1 Quicksilver." Sales under this contract averaged over $110 per flask, although 400 flasks were sold as one block for $90 per flask. Sales were made through Jecker, Torre & Co. of Mazatlán and Barron, Forbes & Co.'s subagent at Guanajuato, Messrs Rohl & Goeome.

6. Ibid., vol. 2, folder 1.

7. Ibid., vol. 2, folder 1. By the third statement Bolton, Barron & Co. were also shipping quicksilver to London, New York, and China. This statement shows 15,925 flasks sold for a gross income of $557,460.

8. Flasks of mercury reported as exported were not necessarily the same flasks as those counted in the reported production in any given year.

9. The New Idria case was the basis for a novel by Bret Harte: *Story of a Mine*.

10. Walter W. Bradley, *Quicksilver Resources of California*, 139. In addition to purchasing the New Idria Mine, Bradley states, Messrs. Barron & Co. also purchased the Josephine (later Little Bonanza Mine) in San Luis Obispo County during the time New Almaden was closed by injunction. Barron & Co. apparently spent a substantial sum developing the mine, although poor prospects and the lifting of the injunction caused it to abandon the effort.

11. Brewer, *Up and Down California in 1860–1864*, 160.

12. Barron and Forbes realized the mine's land title problems early on and hired lawyer and prominent early Californian Henry Wager Halleck (1815–72) as mine manager, a position he held from 1850 to 1856. Halleck had been the secretary of state of California under the provisional governor before statehood, and in 1849 had made a study of California land titles, finding that most were imperfect and fraudulent. This view became the overwhelming American view, whereas under the treaty of Guadalupe Hidalgo the Mexicans in the territories that were ceded to the United States were to have their "property of every kind" "inviolably respected." Rawls and Bean, *California*, 130. The legal maneuverings over the ownership of the New Almaden Mine have been well presented by a number of historians, including Schneider, *Quicksilver*; Milton Lanyon and Laurence Bulmore, *Cinnabar Hills*; Coomes, "From Pooyi to the New Almaden Mercury Mine"; Kenneth M. Johnson, *The New Almaden Quicksilver Mine*; and Milton H. Shutes, *Abraham Lincoln and the New Almaden Mine*. Primary source

material includes John T. Doyle, *A Letter to the President of the United States, in Reply to the Attorney General's Report of the Resolution of the Legislature of the State of California*; James Alexander Forbes and Andrés Castillero, *In the Claim of Andrés Castillero to the Mine of New Almaden*; Reverdy Johnson, Jeremiah S. Black, and Charles Fossatt, *The United States v. Charles Fossatt*; Matthew Hall McAllister and Andrés Castillero, *Andrés Castillero v. The United States, on Appeal from the Decree of the United States Commissioners to Ascertain and Settle the Private Land Claims in the State of California*, Brief; Archibald C. Peachy, Gregory Yale, and United States District Court, *The United States v. Andrés Castillero*; and Quicksilver Mining Company, *Charter and Organization of the Quicksilver Mining Company, with Statement of Location, Extent of Property, & Co.*

13. In 1851 the U.S. government set up a land commission to verify landownership throughout California. This committee met from January 1852 to March 1856, during which time every land title in the state had to be ruled on by the commission.

14. For insight into Butterworth's past, and competing readings of his past, see two obituaries: "Death of S. F. Butterworth," *San Francisco Bulletin* published as *Evening Bulletin* 15, no. 25 (May 6, 1875): 3; and "Death of a Millionaire," *Sun* (Baltimore) 12, no. 342 (May 12, 1875): 3.

15. The New Almaden Company had three possible claims to the land encompassing the mine: Castillero's claim to the mine, Castillero's colonist grant of two square leagues surrounding the mine, and the Berryessa Grant that the New Almaden Company had gained from the Berryessa family. If the courts were to uphold either the mine grant or the colonist grant, or if the courts were to find that the mine was on the Berryessa Grant (the boundaries of this grant were not precisely defined under the Mexican government), then the New Almaden Company would own the mine. The challenges to the New Almaden Company's ownership came from the neighboring Larios Grant for the Rancho Capitancillos. The Larios Grant started at one end of a long mountain ridge that at the other end had the New Almaden Mine. Because valley land was what concerned the people at the time of the original grants, the boundary lines of the grants through the mountains were not well defined. For the parties controlling the Larios Grant to gain the New Almaden Mine, they had to convince the courts to dismiss Castillero's mine grant and colonist grant, and find that the New Almaden Mine was actually part of the Larios Grant. In addition to the New Almaden Company, and the various groups holding the Larios Grant, the government claimed the New Almaden Mine, arguing that neither the Berryessa Grant nor the Larios Grant included the mine, and the mine was therefore on government property.

16. Schneider, *Quicksilver*, 36.

17. The lawyers for the government were former attorney general Jeremiah S. Black, assisted by the current attorney general, Edward Bates, and Benjamin R. Curtis. The latter attorney was paid by the QMC, a fact that did not escape commentary in the press. Coomes, "From Pooyi to the New Almaden Mercury Mine," 112.

18. *U.S. Appellants v. Charles Fossatt*, 1857.

19. Swetl had been central in Lincoln's election. See Shutes, *Abraham Lincoln and the New Almaden Mine*.

20. The writ was based on the Act of March 3, 1807, chap. 46, entitled, "An Act to prevent settlements being made on land ceded to the United States until authorized by law."

21. Shutes, *Abraham Lincoln and the New Almaden Mine*, 10.

22. Thomas Frederick Bell and William Henry Long, "Dictation Concerning the New Almaden Quicksilver Mines."

23. William Barron had a younger brother, Joseph, who was also involved in the dealings with the mine. Eustace II, the son of Eustace Barron, also played a role in the mine, although for the most part he managed the Barron, Forbes & Co. assets in Mexico. William Barron and Thomas Bell—as the two partners of Barron & Co. after 1858—possibly had the greatest influence over the decisions about the New Almaden Company during the legal battles, especially given that Eustace Barron Sr. had died in 1859 and Alexander Forbes had retired to England. For Thomas Bell as the Quicksilver King, see Hubert Howe Bancroft et al., *Biography of William C. Ralston Prepared for Chronicles of the Kings*.

24. Secretary of the Interior Usher appointed Sherman Day, a prominent engineer, as superintendent when control of the mine was transferred in November 1863. James Eldredge was in charge of the business interests of the QMC on the Pacific Coast from that date until April 1, 1864, when Day assumed both positions until Butterworth took over as general agent and factor on July 1, 1864. See the *Annual Report of the Quicksilver Mining Company*, 1864.

25. Charles Fossatt and Quicksilver Mining Company, *Facts Concerning the Quicksilver Mines in Santa Clara County, California*, 6.

26. New Almaden Mine Collection, 1845–1973.

27. At this time the Spanish government did not have a deal with the Rothschilds. For letters detailing the structure of the early QMC sales, see New Almaden Mine Collection, 1845–1973, box 11(1).

28. Also great years of production were 1863 and 1866.

29. Quicksilver Mining Company, *Reports and Exhibits Submitted at the Annual Meeting of the Stockholders* (1864), 40. This report was printed after the annual Board of Directors meeting on February 2, 1865.

30. Ibid.; see the chart on the last page of the report.

31. New Almaden Mine Collection, 1845–1973, box 5, bk. 2, 136. This mercury was probably the 500 flasks per month that Butterworth was shipping to the London Rothschilds as payment on the $250,000 loan that the QMC had taken from August Belmont, the New York agent for the Rothschilds. The Rothschilds and Barron, Forbes & Co. had divided Mexico for the purpose of quicksilver sales into western and eastern Mexico, consistent with the Rothschilds' control of the Atlantic and Barron, Forbes & Co.'s control of the Pacific.

32. The New Almaden Mine records during these years are full of quicksilver sales proposals from the Rothschilds, Barron & Co., and minor houses. In the 1864 QMC Annual Report, the president stated: "The stock of the company in its present low price in gold, has been to some extent purchased for account of parties on the Pacific Coast, and a sufficient quantity has already gone into European hands to bring before your Board of Directors several applications for the establishment of agencies and opening transfer offices in London and upon

the continent. The perfecting of these arrangements is a subject which should receive early attention from the Directors." Quicksilver Mining Company, *Reports and Exhibits Submitted at the Annual Meeting of the Stockholders* (1864), 40. See also New Almaden Mine Collection, 1845–1973, box 11.

33. New Almaden Mine Collection, 1845–1973, box 11.

34. Ibid., box 29, folder 2.

35. In this document, the Barron, Forbes & Co. interests are called Eustace Barron & Company (after the retirement of Forbes). William Barron was to pay thirty dollars in gold per flask, half on delivery and half in ninety days. He was also to advance the QMC $215,000 on their floating debt and $150,000 on interest, reimbursed by the delivery of quicksilver. Additional advances, if required, were also arranged by this contract. Quicksilver Mining Company, *Reports and Exhibits Submitted at the Annual Meeting of the Stockholders* (1866).

36. Interestingly, his letters to his former agents do not name the party to whom he had sold the mercury.

37. D. O. Mills—a prominent and respected banker—was the president of the Bank of California, although Ralston, as treasurer, is acknowledged as the man who managed the bank. William Sharon (later Senator Sharon of Nevada) established branches of this bank at Virginia City and Gold Hill, Nevada, during the first Comstock Boom. During its first years of operation, the Bank of California was immensely profitable, raising the bankers themselves to prominence in San Francisco and California. Thomas Frederick Bell, "Dictation Concerning William Ralston" (2 pages).

38. Bell's death announcement states that he was one of these original incorporators of the bank. "Death in a Stairwell: The Fatal Fall of Capitalist Thomas Bell," *Morning Call*, October 17, 1892.

39. The following prominent histories cover both the Comstock and the Bank Crowd: Grant H. Smith, with new material by Joseph V. Tingley, *The History of the Comstock Lode, 1850–1997*; Ronald M. James, *The Roar and the Silence*; and Rodman W. Paul and Elliott West, *Mining Frontiers of the Far West*.

40. Hubert Howe Bancroft et al., *Biography of William C. Ralston Prepared for Chronicles of the Kings*, n.p.

41. Much of the rhetoric of the Supreme Court case that was in support of the New Almaden Company touted the hard work of the men of the New Almaden Company, who had carved the mine from the wilderness of California, taking a wasteland and developing it into the richest mine in the United States. The New Almaden Company supporters attacked the QMC, calling its members idle and crafty, a "hateful and sweatless class, who are willing to prey upon hoards which they have not gathered." This analysis is from Coomes, "From Pooyi to the New Almaden Mercury Mine," 142.

42. Letter from Ralston to Butterworth: New Almaden Mine Collection, 1845–1973, box 11, folder 4.

43. Apparently this issue was arbitrated, and the QMC lost. See Schneider, *Quicksilver*, 42.

44. Barnes, Ralston, Mills, and Thompson are listed as the trustees of the New Idria Mine in the *San Francisco Bulletin* 21, no. 76 (January 6, 1866): 4. The trustees had an ongoing land title battle with William McGarrahan over the New Idria Mine. In the end the trustees retained their title over the mine.

45. While the mine was discovered as early as 1862, it was reorganized in 1867 as the Redington Quicksilver Mining Company and soon became a major producer. To bring the Redington Mine into the combination, Barron & Co. offered it the exclusive sale of quicksilver in California: "The firm of Redington & Co., of San Francisco, were appointed sub-agents for the sale of Quicksilver in California, and to be paid a commission of 2½ per cent on the amount of such sales." Quicksilver Mining Company, *Reports and Exhibits Submitted at the Annual Meeting of the Stockholders* (1869), 8.

46. Obituary for John Redington, *Chemist & Druggist*, June 7, 1890, 760. For more information on the history of the Redington Mine, see Donald O. Haus, "The Knoxville-Redington Mine."

47. Rossiter W. Raymond, *Mineral Resources of the States and Territories West of the Rocky Mountains*, 10. Raymond provides the following rationale for the combination's production figures: "The Pacific States and Territories require altogether about 1,200 flasks... per month; Mexico and South America, 1,000 flasks each; China, 1,000. The total annual demand (for California's markets) does not exceed 50,000 flasks, the production of which is divided among different companies as follows: New Almaden, 24,000; New Idria, 10,000; and Redington, 10,000."

48. Barron & Co. owned a large stock of quicksilver on world markets at this time.

49. Raymond, *Mineral Resources of the States and Territories West of the Rocky Mountains*, 10.

50. Ibid., 166. In 1869 the production figures for these mines were 16,898 for New Almaden, 10,315 for New Idria, and 5,018 for Redington.

51. On the control exercised by Ralston and the Bank Crowd over mining and politics, see Gray Brechin, *Imperial San Francisco*, 38–44.

52. Raymond, *Statistics of Mines and Mining in the States and Territories West of the Rocky Mountains for the Year 1870*, 16.

53. Ibid., 16.

54. Randol, *Papers Relating to Quicksilver Mining, Ca. 1849–1894*. See scrapbook clippings for Butterworth's obituary.

55. Although the Bank Crowd worked to control and suppress the QMC, the individuals on both sides were very willing to collaborate on other related business interests.

56. The North Bloomfield proved to be the richest hydraulic mine in the West. The mine produced $2.8 million in gold between 1866 and 1884. The mine was eventually closed by litigation against hydraulic mining. Brechin, *Imperial San Francisco*, 59.

57. The year 1870 found William E. Barron and Thomas Bell immensely wealthy and in control of the trade of California's quicksilver. The 1870 manuscript census shows William Barron and Thomas Bell living at the same address, with four servants, including three who

were Swiss, a steward, a valet, and a kitchen man, as well as a French chef. William Barron lists himself with $175,000 in total assets, while Thomas Bell claims $500,000.

58. For information on the Comstock, see Brechin, *Imperial San Francisco*, chap. 1; Smith, *The History of the Comstock Lode 1850–1997*; James, *The Roar and the Silence*; Ronald M. James and C. Elizabeth Raymond, eds., *Comstock Women*; and Paul and West, *Mining Frontiers of the Far West*, chaps. 4 and 5.

59. The dates for the three combinations were April 1, 1868, to April 1, 1870; April 1, 1870, to April 1, 1872; and April 1, 1872, to April 1, 1873.

60. Schneider, *Quicksilver*, 49–88.

61. New Almaden Mine Collection, 1845–1973, box 31, folder 3.

62. David J. St. Clair, "New Almaden and California Quicksilver in the Pacific Rim Economy," 287.

63. Randol, *Papers Relating to Quicksilver Mining, Ca. 1849–1894*, vol. 11, letters 21 and 22.

64. *San Francisco Call*, August 31, 1873. From ibid., vols. 7–10.

65. The history of the quicksilver tariff in the United States is as follows: 1842, 5 percent; 1846, 20 percent; 1857, 15 percent; March 2, 1861, 10 percent; during the Civil War, 15 percent; 1867, about 15 percent, lasting until 1873 when it was effectively reduced by 10 percent; February 8, 1875, no tariff; June 19, 1894, seven cents a pound, renewed in 1911. See Randol, *Papers Relating to Quicksilver Mining, Ca. 1849–1894*, vols. 7–10.

66. J. B. Randol was made manager at New Almaden in 1870. Randol was the nephew of Butterworth, and had been the company accountant in New York. Whereas Butterworth had been a society man in San Francisco, Randol was an on-site manager.

67. Randol, *Papers Relating to Quicksilver Mining, Ca. 1849–1894*, vol. 11, letter 22.

68. Ibid., vol. 11, August 17, 1874. Describing what proved to be two of the larger mines in the history of the industry in the state, the report states that the Great Western Mine had high-grade ore and looked very promising, while the chances of a permanent mine at the Sulphur Bank were not good.

69. "Quicksilver," *Mining and Scientific Press*, January 23, 1875.

70. Although it is not clear when, this 1870 contract was abandoned and the Redington Mine negotiated a deal with the Comstock silver kings James Flood and William O'Brien for 400 flasks a month.

71. In 1875 Daniel Drew, a famous Wall Street speculator as well as the president of the Quicksilver Mining Company, went into bankruptcy. An article in the *San Francisco Bulletin*, March 24, 1876, attributes his failure to heavy losses on Wall Street in 1873 and 1874, losses on the Canadian Southern Railroad, and losses from the QMC. See Randol, *Papers Relating to Quicksilver Mining, Ca. 1849–1894*, scrapbooks.

72. Bell, "Dictation Concerning William Ralston." Bell recounts in this dictation that he also secured credit from the Rothschilds to stabilize the bank.

73. Randol, *Papers Relating to Quicksilver Mining, Ca. 1849–1894*, vol. 11, March 12, 1875.

74. Ibid., vol. 11, March 22, 1875. By early 1876 the price of mercury was at preboom levels, and the price dropped to all-time lows by mid-1878.

75. New Almaden Mine Collection, 1845–1973, box 28, bk. 23, December 14, 1876.

76. Anthony Zellerbach, a paper magnate, owned the Altoona Mine in Trinity County and the Oceanic Mine in San Luis Obispo County. Zellerbach needed a supply of mercury for whitening his paper products. The Parrott family—under the active interest of Tiburcio Parrott, the illegitimate son of John Parrott—controlled the Sulphur Bank Mine in Lake County and had a lease on the Great Eastern Mine in Sonoma County. See Randol, *Papers Relating to Quicksilver Mining, Ca. 1849–1894*, vol. 11, booklet 4.

77. Helen Holdredge claimed that Alvinza Hayward, the president of the Merchant's Exchange Bank, had been a member of the Bank Crowd in the late 1860s, but had been cheated out of Belcher Point Mine stock by William Sharon. Sharon had traded Hayward's Crown Point Mine stock for the Belcher Point stock. However, the Crown Point soon became a valuable mine. Hayward saw he had been cheated and, with the Merchant's Exchange Bank, set up a milling company to rival the Union Mill and Mining Company. See Helen Holdredge, *Mammy Pleasant's Partner*, 208–9.

78. Randol, *Papers Relating to Quicksilver Mining, Ca. 1849–1894*, vol. 11, booklet 5, letter dated January 24, 1877. Further proposed rules included a provision whereby the owners of quicksilver could take and ship quicksilver as an "adventure," trying to sell to their own customers at a higher price. The expenses of the board of three directors were to be paid pro rata on sales.

79. Ibid., vol. 11, booklet 5, letter from Bell to Randol dated January 24, 1877. Bell goes on to write that there is little demand in China in winter until after their New Year, "the weather not permitting any repainting of houses so that there is little sale of vermillion."

80. This organization was called The Eureka Company.

81. "My Combination," as Bell wrote, allotted him 18,000 flasks a year for New Almaden and New Idria combined, and gave the other mines the following quotas: Parrott & Co. (Sulphur Bank and the Great Eastern), 10,000; Redington, 5,000; Great Western, 4,000; Dore, 5,000; Langley, 1,500; and the Merchant's Exchange Bank, 3,000. This combination limited production to 46,500 flasks a year (not counting the odd lots produced by the small mines). Randol, *Papers Relating to Quicksilver Mining, Ca. 1849–1894*, vol. 11, booklet 5, October 29, 1877. Bell wrote in this letter to Randol: "We should deal liberally with the Union Mill and Mining Company—they take our quicksilver."

82. Andrew S. Johnston, "Quicksilver Landscapes."

83. See Brechin, *Imperial San Francisco*, 49–53, for details on this decision and its impacts on the West. The case was *Woodruff v. North Bloomfield Mining Co.*

84. Although Bell had been an investor in the Virginia and Truckee Railroad, he was not a partner with the new "Railroad Kings" such as Crocker, Stanford, Huntington, and Hopkins. He instead used his strength as a businessman of international connections to try to profit in lands where he was the best equipped. One of his major ventures was the backing of the Central American Railroad in Guatemala.

85. Thomas Bell's private life became public in 1884 and 1885 in two trials involving William Sharon, an early member of the Bank Crowd in the 1860s, and Mary Ellen Pleasant, also known

as Mammy Pleasant, Bell's long-time business partner and companion. In 1884 Pleasant took on former senator William Sharon in court. The case of *Sharon v. Sharon* involved a woman, Sarah Althea Hill, who claimed to be Sharon's wife and was seeking alimony. Sharon claimed she was a prostitute and that he was not married to her. Pleasant financially backed Hill in her case. Historians claim that this case was a means of revenge against Sharon by Pleasant and Bell for cheating them out of wealth following the death of Ralston and the scramble over his assets. For more on Mary Ellen Pleasant and these events, see Lynn M. Hudson, *The Making of "Mammy Pleasant,"* 75–76. For more information on Bell and Pleasant, see also the racy and to some degree fictionalized biography of Thomas Bell: Holdredge, *Mammy Pleasant's Partner*.

86. See Robert L. Spude, "Cyanide and the Flood of Gold."

3

A Geography of Mercury Mining in California

In June 1869, E. R. Sampson wrote to the president and directors of the Quicksilver Mining Company in New York concerning his recent visit to the New Idria Mine:

> I took one of the foremen from New Almaden with me and on inspection we agreed that we saw a good mine. The ore is more plentiful than ours, but not so good quality... They have an abundance of ore on hand and could work double the number of hands to advantage upon ore in sight, if they desired—they pay a less price per carga than we do but the men make better wages at the lesser price owning to the ore being so much more plenty and more easily worked... We found a large number of our miners at work there. They work about 300 men and we about 500.[1]

The development of the New Idria Mine, and the way it was worked through much of the nineteenth century, together serve as an example of the complex interplay of many factors, both natural and cultural, shaping the California mercury landscape. These factors include geology and geography, money and power, and technology and business practice. Although prospected in the early 1850s, the mine was not developed on an industrial scale until William Barron, Thomas Bell, and other members of the Bank Crowd purchased it in 1858, when the New Almaden Mine was threatened with closure by court injunction.[2] As was detailed in chapter 2, Barron, Bell, and the Bank Crowd manipulated production at New Idria, increasing or decreasing production as benefited their attempts to control the quicksilver trade.

Although the use of the New Idria Mine as a tool for controlling the quicksilver market was economic and social, the geology of the mine helped make the Bank Crowd's mode of operation possible. According to Sampson, the cinnabar deposit at the New Idria Mine was extensive and relatively easy to mine. Although the ore at New Idria was poorer than at New Almaden, it was plentiful and it lay in predictable formations, eliminating the need for expensive

DOI: 10.5876/9781607322436:c03

FIGURE 3.1 (FOLLOWING PAGE) Geologic Map of the New Idria Quicksilver District, 1916

The ore bodies of the New Idria and San Carlos Mines occurred in the Franciscan formation (horizontal hatch) near its contact with the Panoche formation (diagonal hatch). The New Idria Mine and the town of New Idria are labeled separately on this map—the mine is just below the town. The cinnabar also occurred as a cementing material binding the Franciscan rocks. North is up. (Bradley, *Quicksilver Resources of California Bulletin No. 78*, pl. 12.)

....

prospecting work as was necessary at New Almaden (Figure 3.1). In addition, the ground was firm, requiring little timbering, yet was more easily dug than the rock at New Almaden, and the mine was also dry, requiring neither elaborate pumping systems nor constant vigilance and maintenance. Best of all for the Bank Crowd, the New Idria ore deposits were easily worked: "As the mountain is very bold and precipitous, all their workings are tapped by a series of tunnels, run at different levels at very small expense" (Figure 3.2).[3] The particular geology of the New Idria Mine aided the Bank Crowd in easily increasing or decreasing production, and together the geology and history of the mine show that ideas such as maximum efficiency and full production are not the most useful ideas for understanding the New Idria Mine landscape.

Instead, considering that its owners did not maximize production—nor did they overly invest in its infrastructure despite the great production capacity of the mine—other themes are best for understanding the mine, especially underdevelopment. The infrastructure at the mine was poorly developed, including notoriously bad roads and a deficient reduction plant, of which Sampson wrote, "I do not consider the reduction works as being in good condition, an evidence of which is the fact that many of the employees are suffering from salivation (mercury poisoning), which could not occur if there were no leakage."[4] He went on: "Should they be disposed to expend more money on furnaces and equipment, they could compete favorably with the New Almaden Mine in product and would no doubt come in on equal footing if the combination were continued."[5] This underdevelopment was a conscious choice, in step with the Bank Crowd's plans for controlling the mercury markets. Interestingly, the New Idria Mine went on to be the largest producing quicksilver mine in the Western Hemisphere in the twentieth century, in part because its cinnabar was only modestly exploited in the nineteenth century, despite the plentiful ore and the relative ease of mining it (Figure 3.3).[6] As Sampson stated, "The great difference to my mind between that mine and the New Almaden is that the New Idria is comparatively a virgin mine, whilst ours is not."[7]

This example of the New Idria Mine demonstrates the importance of understanding the interrelations between mines as physical places in the landscape, on the one hand, and mines as tools constructed and manipulated by competing groups of people to

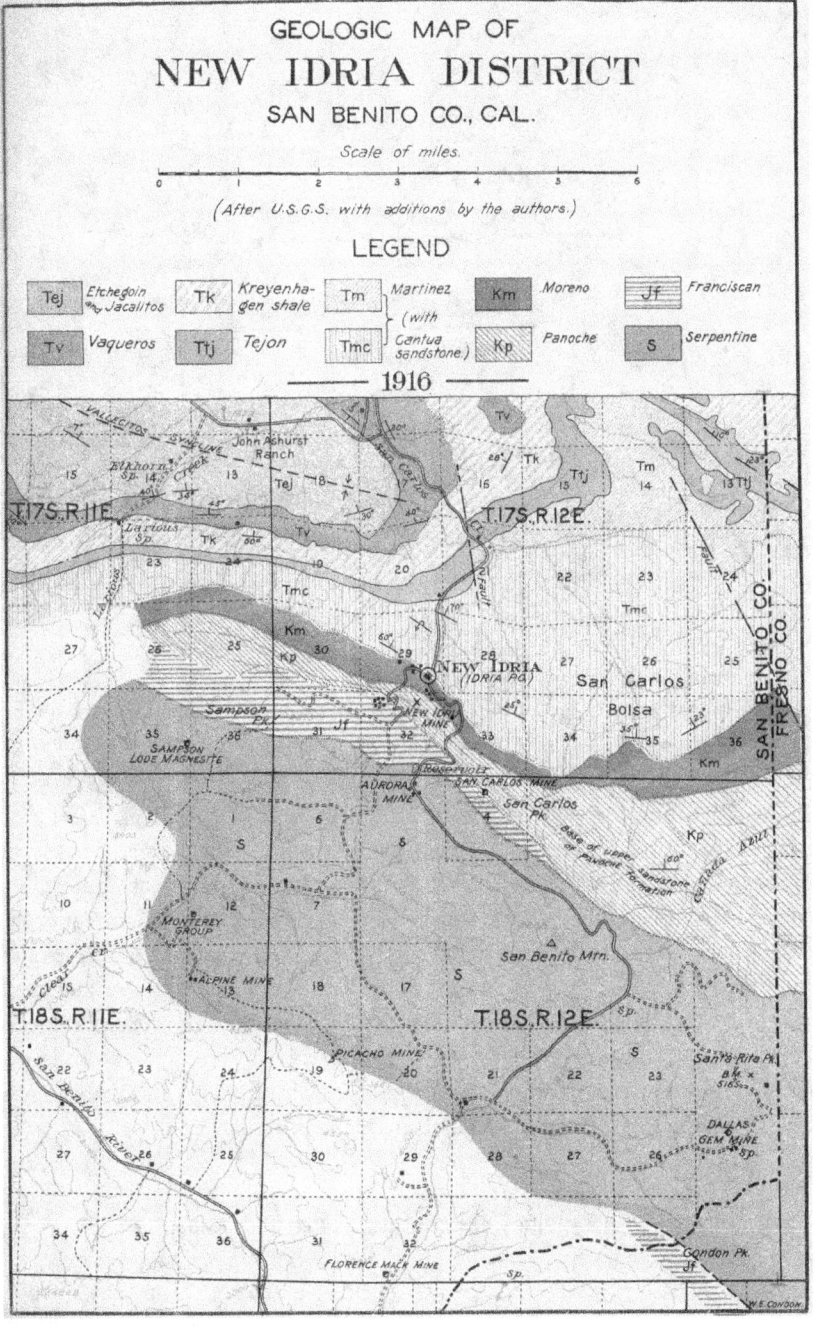

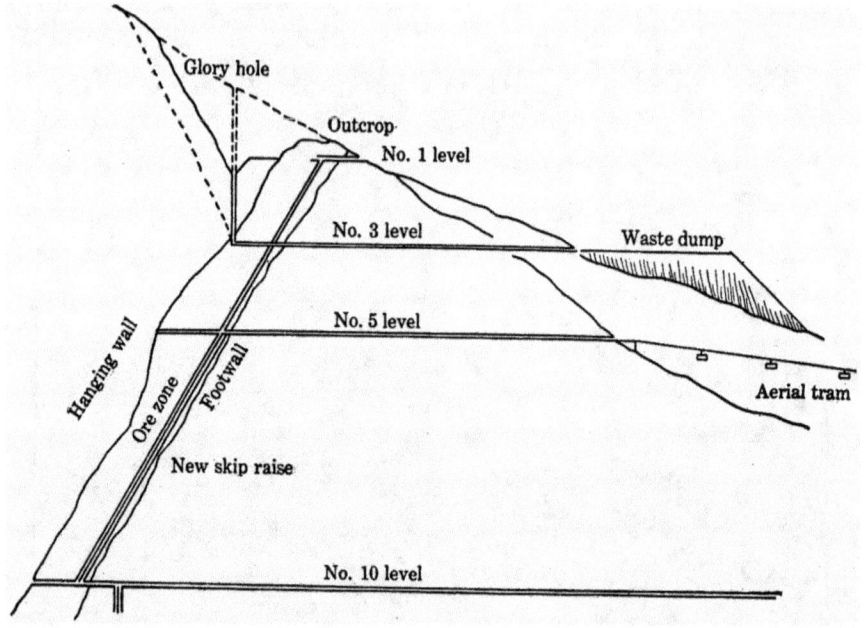

FIGURE 3.2 Section of the New Idria Mine

This drawing shows the primary ore body (labeled "Ore zone") of the New Idria Mine and how it was worked by a series of tunnels coming from the right. The ore body was originally shaped like a lens (here the top of the lens is gone and the profile of the hill has been altered by pit mining). (C. N. Schuette, *Quicksilver*, Bulletin 335, 20.)

serve various ends, on the other. Although geology alone might seem to be the simple explanation of the location and character of the quicksilver mines of California, the mercury mines and camps of California reflect more than simply an efficient exploitation of geology. We cannot learn from geology why some ore deposits were exploited while others were not, or why some mines looked one way while others looked another. A quicksilver mine was not only a cinnabar deposit; it was also the means people used to exploit the cinnabar and gain wealth from it. In this way mines can be understood as tools, created by groups of people to produce wealth from a mineral resource and exhibiting a range of interrelated natural and human complexities.

A Typology of California Quicksilver Mines

The New Almaden Mine, its neighbor on the other side of Mine Hill, the Guadalupe Mine, and the New Idria Mine are the original, higher-grade mercury mines in

FIGURE 3.3 Panoramic View of the New Idria Mine, Plant, and Town about 1918

This view shows New Idria built up to meet high World War I production demands. The New Idria ore body is in the mountain at the center of this photograph. Tunnels were driven to the ore body from various locations and elevations on the north- and east-facing slopes of this mountain. The San Carlos Mine is at the top of the large hill at left in the distance. The view is facing south; east is to the left. (Bradley, *Quicksilver Resources of California Bulletin No. 78*, 108.)

Mine Type	First Developed	Worker Ethnicity	Type of Camp	Ore Type	Initial Furnace
Original Mines	1850s	Mexican/Chilean	Hacienda/Plantation	Higher-grade	Intermittent
Redington Mine	1862	Irish	Company Town	Lower-grade	Large Intermittent
New Mines	Mid-1870s	Chinese	Minimal/Contracted	Lower-grade	Continuous
Boom Districts	Early 1860s, 1870s	White	Speculative Camps	Small Quantities	Retort

FIGURE 3.4 A Typology of California Quicksilver Mines

There were various types of quicksilver mining in California, and these types changed over time. Thomas Bell used the terms *original mines* and *new mines* to categorize quicksilver mines in his correspondence in the 1870s. Worker ethnicity describes the greatest number of laborers when the mine was first developed. This typology of quicksilver mining in California helps in understanding the transitions that occurred in the industry, both at the physical mines and in the practices of business, from its earliest days in the late 1840s until the 1890s. Also, understanding the reasons behind the development of these particular types and how they changed can give us insight into the social, cultural, and physical development of California. (Chart by Austin Porter.)

California. These mines were located by the 1850s, and were developed based on the hacienda model employed by Barron, Forbes & Co. at New Almaden. This hacienda model required significant capital outlay from the owners for the development of the mine, reduction works, and settlement. The workforce at these mines—originally mostly Mexican and Chilean—came over time to be organized according to a complex racial and ethnic hierarchy with the addition of Cornish, Swiss, Chinese, and other ethnic groups. These original mines were located south of San Francisco Bay, with the nearest towns being San Jose in the case of New Almaden and Guadalupe, and Monterey and San Juan in the case of New Idria (see Figure 3.4).[8]

The Redington Mine, with its company town Knoxville, is unique among quicksilver mines in California. Located in a remote valley north of San Francisco Bay, it was an easily worked mine with large deposits of lower-grade ore relatively near the surface. The mine was owned and managed by men based in San Francisco with strong ties to gold and silver mining in California and Nevada, and they developed their quicksilver mine and camp based on examples from these industries. They hired Irish miners and laborers with gold mining backgrounds, and interestingly these miners were the only group of Irish cinnabar miners in California in the nineteenth century. A union hall was built at the mine, so presumably the mine workers were unionized, making them also the only unionized quicksilver miners in nineteenth-century California.[9] Knoxville was a company-owned town with boardinghouses, kitchens, family cottages, and company stores. The Redington Mine and Knoxville together serve as an interesting comparative example, for while the company modeled itself as far as metal production on the gold and silver mines of the day, it also had to compete with the other quicksilver mines in the industry in the mercury marketplace.

In contrast to the original mines and the Redington Mine, the new, lower-grade mines—such as Sulphur Bank, Great Western, and Napa Consolidated (Oat Hill)—were developed during the quicksilver boom years of the 1870s, when the Magic Quicksilver Ring lost its control of the industry. The new mines, reflecting a labor structure pioneered by the transcontinental railroads in the late 1860s, established a workforce with a simple racial and ethnic hierarchy of Chinese miners and laborers paid through a labor broker on a group contract, and wage labor whites who held skilled and supervisory positions.[10] The new mines were mostly north of San Francisco Bay in Napa, Sonoma, and Lake Counties. Other new mines were developed in the 1870s in the far north of California in Trinity County (Altoona and Integral), and in the south in San Luis Obispo County (Oceanic and Klau), but these mines were mostly minor producers during the 1870s. Some of these mines became significant producers in the late nineteenth and early twentieth centuries. There were existing towns near most of the new quicksilver mines, relieving the new mines from having to provide many of the services that the original mines had to provide, making the initial outlay of development capital for the owners lower than that for earlier quicksilver mine types.

The quicksilver boom districts were of minor importance economically to the industry, but are interesting comparatively due to many similarities with gold rush districts, with a similar method of making mining claims and their similar boom camps. The two most prominent boom districts were Pine Flat (Sonoma County) and Sulphur Creek (straddling Lake and Colusa Counties), both north of San Francisco Bay. There were two periods in California when there were quicksilver booms: first, in the early 1860s, when New Almaden was closed by court injunction and there was disarray in the industry and a rise in quicksilver price, and second, early in the 1870s, when again the industry was in turmoil and there was a rise in price. These districts were prospected and owned mostly by local whites, who often developed the mines with modest amounts of capital and their own labor, with the goal of selling the prospects to capitalists. Although a few moderately successful mines emerged from these districts, most prospects were abandoned by 1880.

This typology of mercury mines in California establishes four competing models of quicksilver production that can be compared to explore the industry, its history, and its role in the development of California. What factors led to the emergence of these types? Why were these types both economically viable and socially acceptable? How were the physical and social aspects of these types negotiated and contested by various involved groups? The answers to these questions, explored in the rest of the book, involve understanding the fundamentals of the quicksilver industry in California, beginning with the resource, the ore deposit itself, and continuing to describe the ways people contested and developed the resource.

The Economic Geology of Cinnabar

How mine operators and miners understood the geology of quicksilver was based on practical knowledge gained from actual mining, predictive theories about quicksilver deposition, geologic structure, and guesses. There was no unanimity in geological understanding and theory; rather, different contexts made for different understandings, guided by science, folklore, and practical experience.[11] For prospectors and miners in the nineteenth century, the primary ore of mercury, cinnabar, was rare and occurred in economically viable quantities for mining at only a few sites in the state, and indeed in the world. From their experience mercury, like most metals, was deposited in areas of the globe with a history of episodes of mountain building. As they understood it, metal ores, mercury among them, were deposited in these geologically active regions by hot aqueous solutions rising from deep within the earth through fissures in the rock. The form that mercury deposits took depended on the hardness, porosity, and character of the fracture that was invaded. Often cinnabar deposits occurred as veins as the metal ore filled the rock fissures.

George Becker, a geologist who in 1888 prepared the first major study of the quicksilver resources of California, described many quicksilver deposits as "chambered veins," where the veins varied over their length from only the slightest indication of color to larger pockets of very rich ore, depending on the size and shape of the fissure being filled (Figure 3.5).[12] One common type of fissure was formed through the movement of massive blocks of rock that cracked and separated in broad planes, giving the fissures—and consequently the cinnabar deposits that formed in them—a broad three-dimensional quality across the plane of the fissure. Typically these fissures were not uniformly filled with ore; instead, these deposits formed an irregular, netlike structure in the plane of the fissure. Quicksilver ore was not always found in veins. Cinnabar could also be deposited as the aqueous solutions carrying it penetrated and filled porous rock. In the quicksilver mines of northern California, this was often the case with sandstones that were impregnated with cinnabar. Mercury could also be deposited as replacement ore, where the aqueous solutions dissolved the existing rock and replaced it with the mineral.

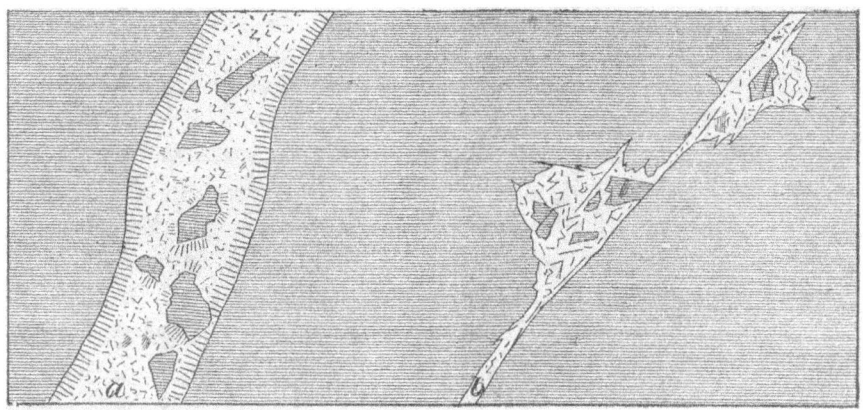

FIGURE 3.5 Drawing of Two Types of Quicksilver Veins
The simple fissure vein (left) and the chambered vein (right) are two types of quicksilver veins identified by geologist George Becker in his 1888 study. Cinnabar is denoted by the horizontally hatched chunks in the veins. Scale is variable. (Becker, *Geology of the Quicksilver Deposits of the Pacific Slope*, fig. 20.)

Practical experience had shown that quicksilver ore bodies were irregular, making the job of mining cinnabar particularly difficult. Even compared with gold and silver mining, where the ore was also deposited in veins, mercury mining was particularly problematic due to the unpredictable structure of the ore deposits. Miners and geologists often attributed human characteristics to cinnabar, referring to it as capricious and impulsive as well as unpredictable. The largest mines, such as the New Almaden, were closely studied by geologists and mining engineers, and mining proceeded according to the best judgment of these experts based on their mine surveys, research, and theories. But despite this level of expertise, even the largest mines showed high variability in month-to-month production due to the unpredictable nature of the ore. George Becker, the California state geologist, warned quicksilver miners in 1888 regarding the unpredictable quality of quicksilver ores: "The special attention of superintendents should be directed to a study of the fissure system. This will almost invariably be complex, and can satisfactorily be made out only by daily study as the work progresses."[13] At smaller mines, where formal expertise was nearly always lacking, miners sought ore based on what they learned from practical experience. "Cinnabar spouts like the whale" was one rule miners used to understand the structure of ore deposits as they worked in the mines. While describing a structure, this bit of practical wisdom was also a theory on the deposition of cinnabar, that is, how the deposits were geologically formed in the first place.

Mining engineer and geologist C. N. Schuette published an article in 1937 describing his theories on the formation of cinnabar deposits, based on his years of practical experience studying and working in most of the California and Texas quicksilver mines.[14] His theories are those of a practical man familiar with the mines whose job it was to understand the deposits for economic gain. Given that he visited or worked at most of the quicksilver mines in the West, his article is a synthesis of the available practical and theoretical knowledge for understanding quicksilver ore deposits. Like most experts Schuette accepted the theory that the source of mercury was deep-seated magma, and that the hot alkaline solutions ascending from this magma through fissures in the rocks carried dissolved minerals until conditions existed where they were deposited. Schuette's theories follow from those of geologist Samuel Christy, who in 1878 wrote an article asserting that cinnabar was deposited by hot, alkaline solutions as they were forced toward the surface.[15] Schuette's contribution, based on his practical experience at the mines, was to identify the common geologic structure necessary for the initial formation of mercury deposits. Figure 3.6, from Schuette's article, details the geologic structure characteristic of the California quicksilver mines as he understood them. He found that the ore bodies formed where the hot alkaline solutions rising in fissures were blocked beneath layers of impervious rock. Once trapped, the ore minerals then precipitated, or condensed, out of the solution due to a combination of factors such as cooling, dilution, a loss of pressure, or contact with agents or reagents that caused precipitation. Schuette recognized that the layers of impervious rock took many forms at the many mines in California, Texas, and elsewhere.[16] This basic insight was valuable information that could be used by practical mining men, and that in fact had probably been understood by experienced quicksilver miners in the local conditions of the mines where they worked.

A big question Schuette sought to answer was why the California quicksilver region was so large, extending nearly the length of the Coast Ranges, from Santa Barbara to the Oregon border (Figure 0.4). Elsewhere in the world, cinnabar deposits were localized—such as at Almadén, Idrija, or Huancavelica—or in a tightly defined region, such as in the Big Bend of Texas.[17] His straightforward answer, again that of a practical man, was that in California geologic events produced the perfect structure for the formation of quicksilver deposits over a large area. In the Coast Ranges of California sedimentary layers of sandstones and shales were subject to various eras of upheaval that included faulting, uplift, and lava flows.[18] The resulting jumble produced many locations where fissures led to areas of impregnated sandstone capped by impervious basalt and shales. Under ideal conditions, when cinnabar rose through these faults, the hot aqueous solutions were trapped in sandstone and formed cinnabar ore bodies.

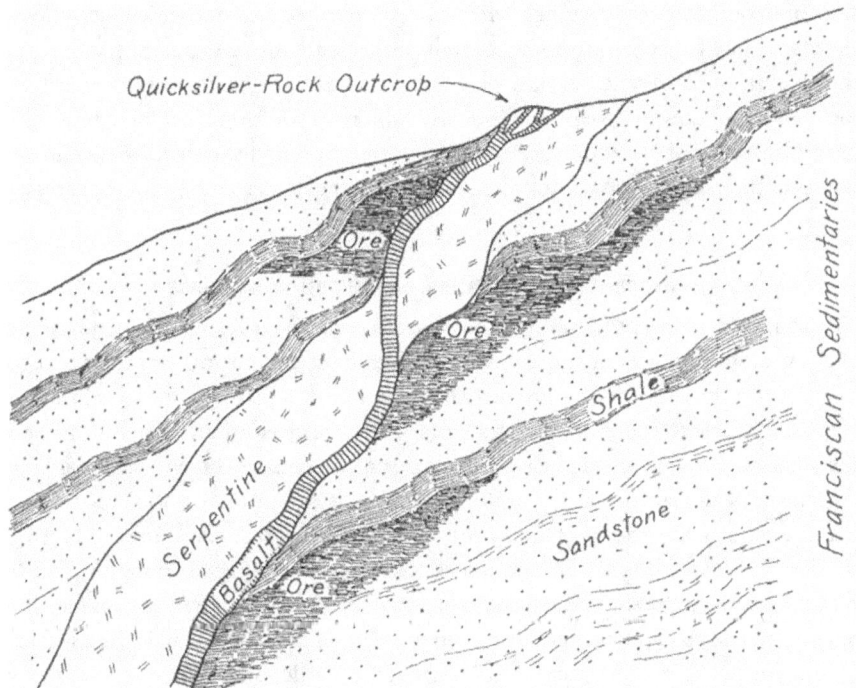

FIGURE 3.6 Idealized Section through a Typical California Quicksilver Mine
The country rocks (rocks characteristic to the area) are sedimentary Franciscan sandstones and shales. A sill (igneous rock injected while molten) of serpentine cuts across the stratified Franciscan layers, while a later dike of basalt is shown coming up along a plane of weakness on the serpentine contact. Cinnabar (labeled "Ore") was deposited by hot aqueous solutions that became trapped under the impervious layers of shale while forcing their way up along the basalt dike. (Schuette, "The Geology of Quicksilver Ore Deposits," fig. 1.)

For the most part the known California mercury deposits tended to be found near the surface, both because this was where the precipitation of the ore primarily occurred and because prospectors could find only deposits that were near the surface. Rarely were working depths at California mines more than 600 feet, and most of the richest deposits were above 300 feet.[19] Some cinnabar deposits, generally those north of San Francisco Bay, were geologically recent and associated with younger volcanic rocks. At mines in the Clear Lake region of northern California and in the thermal regions of Sonoma County, geologists reported the active deposition of quicksilver.[20] The mines in these areas often had intact ore bodies in their original state of deposition, meaning that while still complicated and unpredictable, at least they were not faulted and shifted as well. Other quicksilver deposits, particularly those south of San Francisco Bay, were older, the result of much earlier deposition forming ore bodies that were subsequently subject to additional geologic processes. New Almaden and New Idria are examples of this type, and the structure of the ore bodies at these mines was complicated by the action of these later geologic processes.

Prospecting, Promoting, and Suppressing Mines

Cinnabar deposits were not located by rigorous scientific investigation. Knowledge of the geologic theories of cinnabar deposition (in practical form) was of benefit to the prospectors, but was of more use when developing a prospect into a mine. Not surprisingly, due to the character of the ore, the quicksilver mines of California and the West were found mostly by chance, or by prospectors who spent years roaming the landscape of California's Coast Ranges exploring every outcrop and every ravine. Because the structure of quicksilver ore bodies demanded that the rising ore-bearing solutions be trapped underground, quicksilver ore deposits rarely outcrop; that is, they do not reach the surface, meaning that other geologic processes, such as faulting or erosion, had to happen after deposition to make an ore body visible at the surface. Adding further to the difficulty of locating quicksilver deposits is the fact that many forms of cinnabar rapidly degrade when exposed to oxygen and water.

People looked for cinnabar by association; it was often found in association with certain rock formations and certain minerals, and prospectors learned to use these relationships to increase their chances of locating ore. The California state rock, serpentine, was a good sign, for cinnabar deposits were often associated with serpentine outcrops. Serpentine is a distinctive shiny green rock with a polished surface, and outcrops of the rock were recognizable to the trained eye even at a distance by a distinctive soil color, a lack of grasses, and Digger pine trees. Geologically, serpentine deposits identified faults, faults where cinnabar might have been deposited. Prospectors could also look for deposits of sulphur, as cinnabar is HgS, mercury sulfide. Not all deposits of sulphur have cinnabar, but all deposits of cinnabar have sulphur. A prospector could literally sniff out sulphur hot springs, of which there are many in the Coast Ranges. These hot springs, caused by hot aqueous solutions rising from deep in the ground, were often associated with the deposition of cinnabar. Once zeroed in on a promising area, prospectors looked closely for layers of impervious rock that may have trapped cinnabar beneath it.

A cinnabar prospect was like any other precious metal prospect; when it was located, the owners tried to promote it with the hopes of selling it. Local newspapers often wrote stories on the prospects in their role as promoters of their town and county. On June 11, 1874, the Santa Barbara County newspaper the *Index* promoted a prospect:

> There is a great excitement in certain moneyed and mining rings in San Francisco over reports that have reached there in regard to the discovery of an immense ledge of cinnabar in Santa Ynez Valley, about seven miles north of Santa Barbara ... Mr. Brown (a geologist from San Francisco) is acquainted with every well-known cinnabar mine or channel in the United States and Mexico, and freely gives it as his opinion that there never was before discovered a cinnabar channel that showed such a great extent of ore on the surface.[21]

Such optimistic language was commonplace and part of the promotion game. This particular report came in 1874 during the quicksilver boom, and the locators may or may not have made money selling it to "San Francisco capitalists." Although mining work was done in the area described in this report, few flasks of quicksilver were ever produced.[22]

Often people wanted to develop or "prove" their prospect either with the dream of developing it themselves or with the hope of increasing its value to prospective purchasers.[23] Proving the prospect meant showing ore and producing mercury, but developing a prospect was risky, because you could also disprove the prospect. Developing a prospect also cost money. Although many men, and a few women, developed mercury prospects with their own labor, few of these mines amounted to much. Some small-time miners lived with the belief that they could get rich from their mine, but developing a prospect into an efficient large mine could not be done without significant working capital. The best that small-time operators could do, if they had a rich prospect and wanted to hold on to it to some extent, was to sell interest in their prospect or lease the mine to be paid in a percentage of the production.[24]

Most quicksilver prospects were sold to a group of investors, often local groups from a neighboring town if the prospect was modest, or capitalists from San Francisco for the most promising prospects. Prospects were sold to eastern investors, but few if any of these ever became significant production mines. A few prospects were also sold to British investors anxious to profit from the mining wealth of the American West. Again, few if any British investors made money in California quicksilver.[25] Once investors purchased the mine, they could develop it, try to resell it, or simply hold on to it waiting for a better opportunity to develop it or sell it while keeping it off the market and out of another group's hands. Locating and promoting quicksilver mine prospects occurred mainly during the boom periods of the industry. The rest of the time, the quicksilver powers that dominated the industry either suppressed prospecting or used strong-arm tactics to purchase all prospects.

Although the New Almaden Company's dominance of California quicksilver was nearly total throughout most of the 1850s, this did not stop people from trying to develop other quicksilver prospects. From very early in its history, however, the control of land in the mercury mining industry was established differently from the control of land in gold mining, and this difference resulted in quicksilver prospects being relatively easy to suppress for powerful interests. Establishing mercury mining as different began on December 4, 1854, at a miners' convention held in the "San Carlos District in the Counties of Monterey and Mariposa," later renamed the New Idria District.[26] Based on the proceedings from the event, this miners' convention was held for the same reason as were conventions in gold districts, to establish a mining district and to set the rules for that district, primarily how claims were to be made.[27] In the proceedings it was argued that quicksilver and silver mining were different from gold mining, and because of the difference should be subject to a different set of rules from those commonly adopted in gold districts. The San Carlos proceedings stated:

> Inasmuch as the prosecution of Quicksilver & Silver Mining, require a larger outlay and expenditure of capital than is required, and facilities not requisite in mining for gold, we hereby declare as the law of this District that the discoverers heretofore or hereafter of all Quicksilver or Silver Mines, shall be entitled to the absolute property in said mine or mines upon filing in the office of the Clerk of Mariposa or Monterey Counties, a description of the boundaries thereof, and said mines or any portion thereof, may be sold or transferred by Bill of Sale or Mortgage or Deed, in like manner as other property within the State, hereby confirming all transfers heretofore made.[28]

So unlike those filing a gold claim who had to continually work it to keep up the claim, quicksilver mine locators could file a claim and then it was theirs. This document by Pitts and his colleagues was a radical attempt to rewrite the rules concerning mining claims and property in early California based on an argument that quicksilver and silver mining were different from placer gold mining in that they required much more investment. It does not seem that Pitts and his associates at the San Carlos Miners' Convention had much lasting effect on the district. The later court battles over the ownership of the New Idria Mine, like the battle over the ownership of the New Almaden Mine, hinged on Mexican land grants that preceded any mining claims made after statehood.[29] Pitts's convention may or may not have been a coming together of democratic miners interested in frontier self-governance; what it did result in, however, was New Idria and its many associated prospects being neatly packaged for sale to California capitalists, such as Barron & Co. and its Bank Crowd partners.

Interestingly, although most quicksilver mines were owned outright, during the boom years of the mid-1870s—and to an extent in the late 1850s, while New Almaden was closed by injunction—quicksilver prospectors did set up a few quicksilver districts in the same way gold districts were established. The most prominent new district to arise during the injunction was the Pine Flat or Cinnabar Mining District in the Geysers area of Sonoma County (Figure 3.7). One of the founders of Geyserville, Colonel A. C. Godwin, organized a mining district in 1859, and by 1861 over 33,000 feet of claims had been located in the district.[30] That claims had been made based on tracts of land measured by linear feet meant that the district was organized similarly to placer gold mining districts. Interestingly, this district—located along Sulphur Creek—was one of the rare locations in the state where native, or pure, mercury existed, and where mercury could be panned from the creek bed. This fact of being able to pan for pure mercury, just as gold miners could pan for placer gold, may well have resulted in these districts being established similarly to placer gold districts. Unfortunately for the Cinnabar Mining District, the native mercury was not extensive, and the miners knew little of reducing the cinnabar ore. Most critical to the demise of the district was the reemergence of mercury onto the market from the New Almaden Mine following the injunction. With New Almaden again producing vast quantities of mercury at low cost, the myriad small mines had little chance of competing.

Quicksilver Mining Districts

Prospecting by association was the best tool that people had in locating bodies of cinnabar. The geology of the Coast Ranges was such that once a quicksilver prospect was located, people knew that their best chance for finding ore was in looking nearby for other similar prospects. This relational quality of quicksilver prospects, due to geologic conditions in the Coast Ranges, led to the definition of quicksilver districts throughout the state. The ore formation in a mining district could range from a single isolated ore concentration exploited by one mine—as at the Sulphur Bank Mine on Clear Lake in Lake County—to many ore concentrations associated with a large geologic formation that could stretch for miles and was exploited by many mines, as in the Mayacamas District transecting Napa, Sonoma, and Lake Counties (see Figure 3.7).

Although based on geology, mining districts were also socially and culturally constructed. Each was distinctive and also part of a continuum ranging from the New Almaden District, where one large company dominated, to the Sulphur Creek District in northern California, where there were many small prospects and a couple of moderate-sized mines. The individual mines within a district often had important similarities

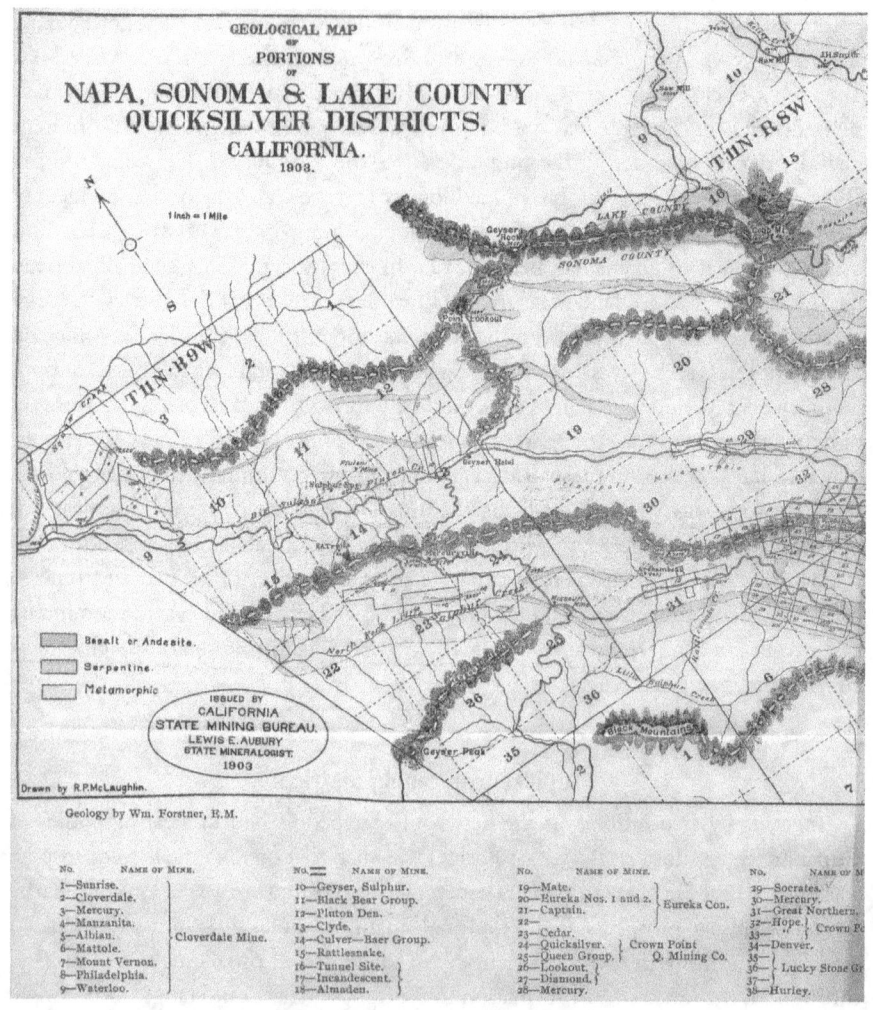

FIGURE 3.7 Geologic Map of the Pine Flat Area, Mayacamas Quicksilver District, 1903

This map details the mines and mine claims (shown as four-sided figures with numbers) of the Pine Flat area (commonly known as the Geysers), comprising the northwest end of the Mayacamas District, which stretched from Cloverdale in Sonoma County to Aetna Springs in Napa County. Shown in this map is about a quarter of the total district, one of the largest. In this geologic map long, wormlike shapes of serpentine lie in a general metamorphic area, the formation characteristic of this district. Cloverdale (Sonoma County) is a few miles to the west from the left edge of the map; Mt. St. Helena (Napa County) is not far off the top right corner of the map. Of the thirty-eight claims listed above, three developed into significant mines. (Aubury, *Quicksilver Resources of California*, after 34.)

> **North of San Francisco Bay**
> Mayacamas District
> Clear Lake District
> Sulphur Creek District
> Knoxville District
>
> **Far North (not on Bradley's list)**
> Trinity Group
> Siskiyou Group
>
> **South of San Francisco Bay**
> New Almaden District
> New Idria District
> San Luis Obispo District
> Los Prietos District

FIGURE 3.8 The California Quicksilver Districts (Bradley, *Quicksilver Resources of California Bulletin No. 78*. Drawn by Austin Porter.)

to the other mines of their district, due to their being developed at the same time to exploit ore under similar conditions. Mining districts were usually developed rapidly following the discovery of an ore formation during a mining boom, but they were also subject to successive waves of development. Individual mines that were developed at one time frequently fell into disuse, only to be reopened later. Quicksilver booms also often saw new mines developed in an old district.

Although discovered by prospectors and developed by mining companies, mining districts were formalized by California state economic geologists who conducted field surveys and then published the results in state mining reports—documents that promoted mining interests and the development of mining.[31] By the mid-1870s the state of California had defined four mercury mining districts south of San Francisco Bay and four districts north of the bay, for a total of eight mercury mining districts in the state, complemented by a few unassociated mines elsewhere in the Coast Ranges (Figure 3.8). These districts were often named after the most prominent member mine. The first district was the New Almaden District (late 1840s), followed closely by the New Idria District (mid-1850s). The 1859–61 boom resulted in lots of claims but little development in a number of fledgling districts, most prominently Pine Flat (called the Cinnabar Mining District at this time, later part of the Mayacamas District). No other new quicksilver districts were defined statewide in the official literature until the boom of the 1870s, except for the Knoxville District and the Redington Mine in 1862. Eventually, these eight districts were how the profession of mining defined the landscape of the industry. County and city boundaries mattered for the taxation of the industry and the voting of the industries' population, but the geology of cinnabar was what mattered for defining the quicksilver districts.[32]

Geologically the districts had distinct characters. The oldest districts—New Almaden, Guadalupe, and New Idria—were characterized by richer ore bodies (in the case of New Almaden, incredibly rich ore bodies) near the surface and easily worked at first. The Knoxville District, discovered while cutting a roadbed, included three principal veins of ore enclosed in serpentine. Less rich than New Almaden (Redington ore averaged 5% mercury), the ore was nevertheless plentiful and near the surface. Unique to the site was cinnabar in acicular, or needlelike, crystal form. The mineral form may have been related to the sulphurous waters and gases present at the site, and that resulted in trouble through corrosion with the reduction plant and machinery.

For the most part, the mines of the new districts established in the 1870s were based on low-grade ore bodies, many of which had been located by prospectors years before but that had no economic value until the boom. In 1873 there had been five producing mines in the state, but by 1876 there were twenty-seven producing mines. Most of these new mines were modest concerns producing at best a few hundred flasks before either the ore or the financial backers gave out. A few of these new mines, however, produced thousands of flasks, and together all of the new mines rapidly came to produce 20 to 40 percent of all mercury produced in the state. Of special note were the Great Western and Napa Consolidated Mines of the Mayacamas District. Each mine operated for thirty years or more. Industrialists also were central to the creation of two new districts in the 1870s. The Clear Lake District was dominated by one mine—the Sulphur Bank—developed by the Parrott family and their trading house. Similarly, industrialist Anthony Zellerbach, another wealthy individual looking both to profit from the quicksilver boom and to secure himself a supply of mercury for use in paper manufacturing, developed the major mines in both the Trinity Group in northern California and the San Luis Obispo District in Southern California.[33] At the other end of the spectrum was the Sulphur Creek District, on the border between Colusa and Lake Counties. This district was not dominated by big capital and one major mine, but instead had multiple owners developing small- and moderate-sized mines in a manner similar to gold rush districts a quarter century earlier.

During the quicksilver boom there was a fleeting moment when the "small man" could enter the quicksilver industry. Mercury being a rich man's metal, the trend had always been big money, major mines, and one dominant player per district. But

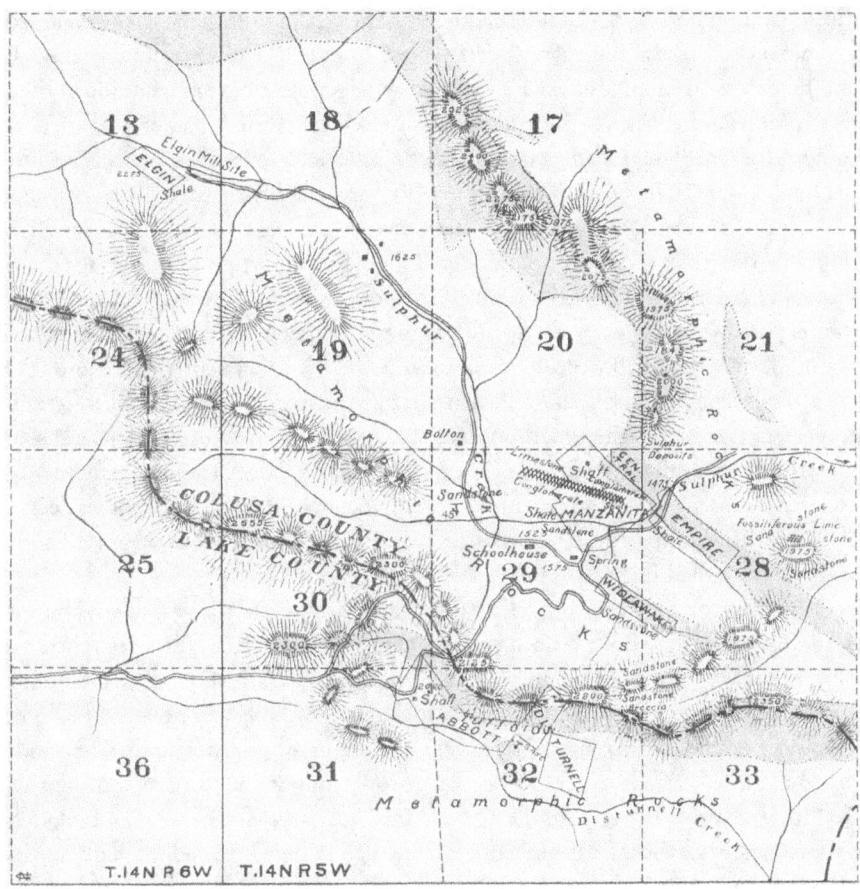

FIGURE 3.9 Geologic Map of the Sulphur Creek District Showing the Disturnells' Claims

The Sulphur Creek District, developed during the quicksilver boom of the 1870s, had at least seven mines that produced significant quantities of quicksilver. Nathaniel and Richard Disturnell worked the Excelsior and other claims on the Lake County side of the ridge running through the Sulphur Creek District (see Section 32). Nathaniel Disturnell's daybook covers their two years living and working in this district. The settlement of Wilbur Springs is just off to the right of Section 28. The Abbott was the largest producing mine in the district. The Manzanita Mine in Section 29 is renowned for having both cinnabar and gold present in the same ore. (Bradley, *Quicksilver Resources of California Bulletin No. 78*, pl. 7.)

during the boom small-time developers got into mercury mining, and they developed a number of districts. These developers ranged from local elites who financed small and moderate-sized mines to sweat-equity partnerships where men developed their own mines. For five years in the midst of the quicksilver production boom, from 1874 to 1878, the "small man" had an opportunity to own and develop a quicksilver mine.

One example is the Disturnell brothers, Nathaniel and Richard, who developed a few prospects in the Sulphur Creek District seventy-five miles north of San Francisco (Figure 3.9).[34] The quicksilver boom offered them the opportunity to work for themselves and eventually even hire a few men to work for them. The mercury prospects they claimed were typical of many of the quicksilver deposits developed during the boom. They were small deposits near the surface with tantalizingly rich ore; the Disturnells reported that their ore, when carefully sorted, contained between 20 and 40 percent mercury.[35] Their methods were pragmatic; the men dug simple and shallow shafts and went after just "the color," the ore itself, not having the resources to concern themselves with the systematic development of theoretical ore bodies deeper in the earth. With only mule power these men then had to lift the ore from the shaft and pack it to their furnace.

Originally the brothers took their ore to a nearby mine to be reduced, but as money came in they built a simple retort (Figure 3.10). Retort furnaces relied on rich ore; it was grossly uneconomical to charge a retort, which could handle only 200–300 pounds of ore, when the ore was low-grade. Although 20 percent ore might yield sixty pounds of mercury per charge, 2 percent ore would yield only six pounds, not enough to warrant the firewood that it took to heat the furnace and the two days' time that it took to take the furnace through a cycle of heating and cooling. Unfortunately for most small-time developers such as the Disturnells, their deposits of rich ore near the surface were rapidly exhausted. If there was more ore present, it tended to be lower-grade ore, the leads of which headed deep into the earth. The lower-grade ore required the new furnaces, which were beyond the means of small-time developers. If these ore leads led to richer deposits deeper in the earth, there was no way for the small-time miners to develop them. If their prospects were at all promising, small-time miners eventually sold their claims to developers with more money; the Disturnells, in fact, eventually sold theirs to the Abbott Mine, the leading producer in the district.

Quicksilver Mines and Camps

The landscape of a quicksilver mine and camp, at its most basic, included a source of ore, a production plant for reducing the ore, a road to the often remote site, and

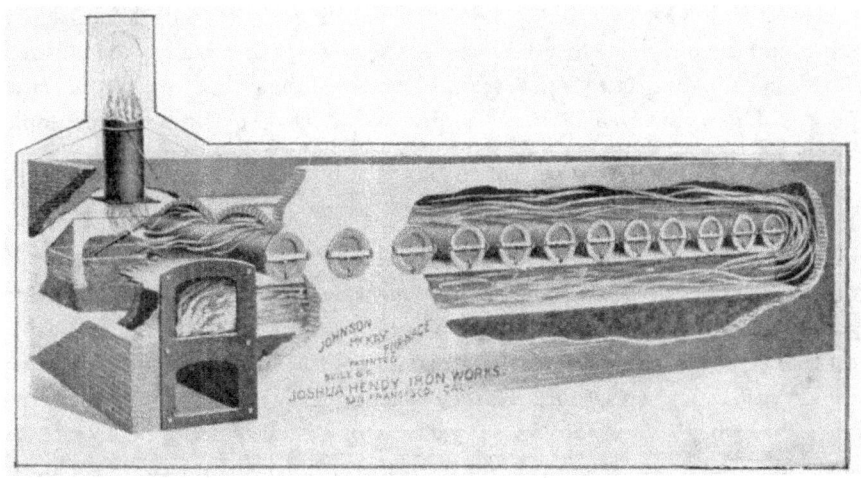

FIGURE 3.10 Johnson and McKay Retort Furnace

This is a twelve-pipe retort patented by Johnson and McKay, and is more elaborate than the pipe retort that the Disturnells would have built. In a retort, the ore is loaded into the pipes and the pipes are sealed. The retort is then fired for a number of hours or days, according to the furnace design and the type of ore, releasing the mercury from the ore as a gas. The retort is then allowed to cool until the mercury condenses. Due to the cycles of firing and cooling, retorts are termed intermittent furnaces. In a retort the products of combustion, soot and ash, are kept separate from the ore. In continuous furnaces, soot and ash mix with the ore. The resultant mercury gas requires an additional heating in a soot furnace to produce "clean" mercury for the market. (Bradley, *Quicksilver Resources of California No. 78*, 213.)

a mining camp that included the basic necessities for the maintenance of the workforce at the mine. The landscape could also included the sources of raw materials for operation—such as wood and water—and supplies for mining and daily camp life.[36] The mines and camps of the quicksilver districts of California were tools created to exploit the mineral wealth of cinnabar. The form and function of the mines and camps varied over time and place and were the result of power struggles among groups of investors, managers, workers, the local community, and the government.

A mine included both underground and aboveground components. Underground were the ore body and the mine that workers and managers built to exploit it; aboveground were the reduction plant, mine offices, and the supplies of resources the miners consumed in operating the mine and plant. The mining camp included the basic necessities for the maintenance of the workers and managers and their families at the mine, including structures for eating, sleeping, and other physical needs, as well as structures for social life at the mines, including religion and education. The mines and camps, while often in remote areas, were not the isolated, preindustrial landscapes often associated with development on the frontier, but were instead dense pockets akin to urbanized industrial landscapes.[37] At a large quicksilver mine people from many countries and of many social classes lived and worked in close proximity.

Quicksilver mines, like most mines, were developed in fits and starts, the products of economic and social incentives to build rapidly, but then abandoned when economics dictated. The cycles of these booms and busts are evident in the landscapes of the mines (Figure 3.11). In boom times mine managers may have decided to construct a new reduction plant or to develop a new mine shaft in their desire to increase production in order to take advantage of the price boom. Most often, a time-consuming project such as a new plant or a new shaft never reached full development while the price was high. These new projects were instead operated in a declining atmosphere where they never were completely finished, and at times were quickly abandoned. However, the time and place where a mine and camp were established mattered. Over time the mine and camp would adapt to change through modifications, but the original structure set in place when first developed was nearly impossible to erase.

The story of the technology of the quicksilver industry, both belowground and aboveground, is a story of power struggles among groups of people involved with the industry and of technological sufficing. The quicksilver industry was not a leader in innovative technology in hardrock mining, nor did the industry as a whole achieve great efficiency in reduction technology. For the most part quicksilver mines tended to follow established mining and reduction practices while operating on smaller budgets than the premier gold or silver mines of the era. The New Almaden Mine is the only exception. It was the most technologically advanced mine in the state for much of the 1850s and remained a marvel for decades, although some California and Nevada gold and silver mines were renowned for greater innovation in the 1860s and 1870s. Especially impressive were the scale of the New Almaden Mine's underground workings and the scale of the reduction yard, with eight separate furnace and condenser systems. Many renowned mining engineers worked at the New Almaden Mine, and both brought ideas to and took ideas away from the operations of the mine.[38] The New Almaden Mine was not, however, a model of technological or scientific efficiency in either mining or reducing ore, nor were many innovative mining or reduction techniques pioneered there. Unlike gold or silver mines, which had to compete in efficiency and in technological innovation with similar mines globally, the mercury mines of California were operated within the context of the tightly controlled California mercury industry, an industry focused on maintaining the status quo and reconciling its own internal logic.

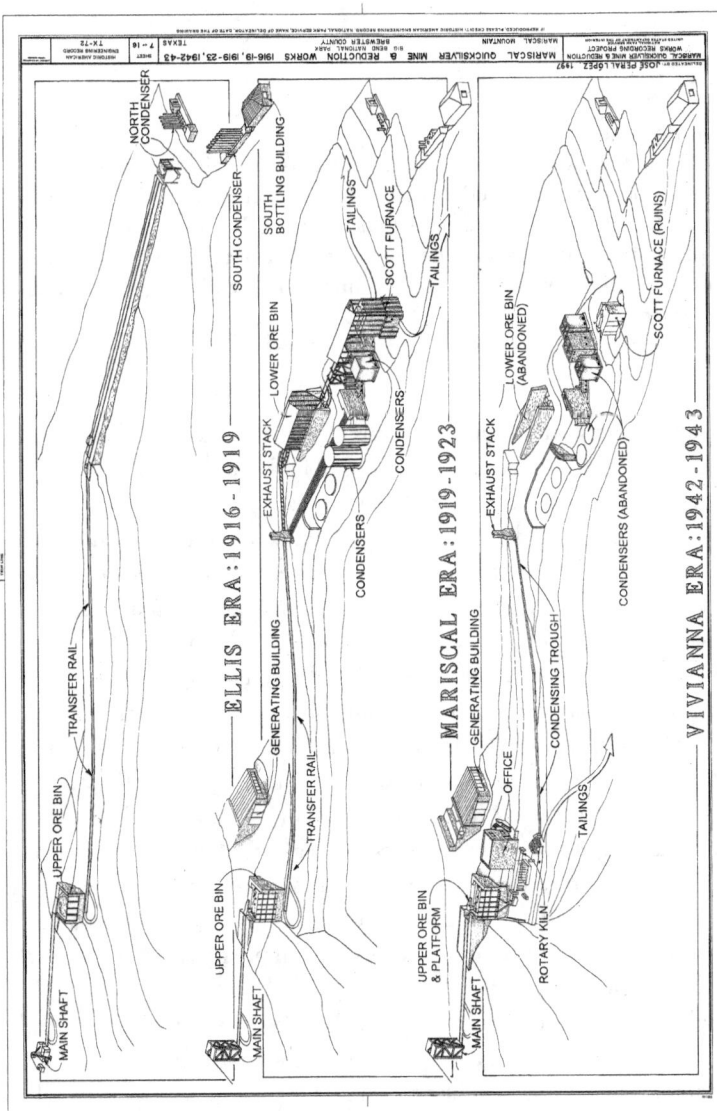

FIGURE 3.11 Three Eras of Mercury Mining Activity in Texas

The Mariscal Mine in Big Bend National Park in Texas had three distinct eras of mining activity. Each era had its own reduction technology, from retorts in the Ellis Era, to a Scott furnace in the Mariscal Era, to a rotary kiln in the Vivianna Era. C. N. Schuette worked at this mine during the Mariscal Era. (Library of Congress, Prints and Photographs Division, Historic American Engineering Record, National Park Service, Jose Peral Lopez, 1997.)

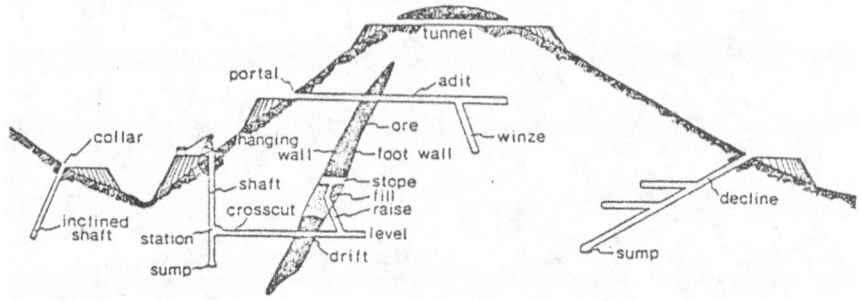

FIGURE 3.12 Anatomy of a Mine

Shown is a sketch of the most common mining elements that miners construct underground in their pursuit of ore. In the center of the sketch is an ore body, slightly tilting from lower left to upper right. From the perspective of the miner working in the ore body, the footwall is the boundary of the ore body underfoot; the hanging wall is the boundary of the ore body overhead. (Intermountain Forest and Range Experiment Station, Forest Service, USDA, Ogden, Utah.)

In the quicksilver mining landscapes, the focus of human endeavor—the greatest effort in reshaping the earth—occurred underground in the mines. It is unfortunate that the greatest work of the miners is largely unseen, being both difficult to access and dangerous, or inaccessible and largely destroyed. Unlike aboveground landscapes, which are created by "positive" built forms in open space, a mine is "negative" space hollowed out from a solid. Buildings aboveground shelter people and their activities from the elements, and similarly the shafts and tunnels of mines shelter humans enough that they can accomplish their work. But unlike buildings, which are built and then remain in one place, the mine is a dynamic shelter. The historian must also confront the fact that until a mine was abandoned, it was in a constant state of adaptation and change. A miner's work involved continual modification of the mine by extending, enlarging, or backfilling tunnels and shafts (Figure 3.12). Unlike many landscapes a historian may study, a mine never reached a state of competition such as a building might when construction ends; the process of mining involved constantly reconfiguring space in the effort to extract ore. The quicksilver industry was based in wresting cinnabar from the earth, and the mines were temporary structures existing only as long as the miners were successful.

At their simplest, mines were scratches or small pokes into the earth, disturbing the earth enough for miners to peek below the surface. At their most complex, mines were elaborate structures resembling upside-down cities, as the geographer

Gray Brechin has argued.[39] Brechin likens the mines of the Comstock, built with a timbering method called the "square set," to the monumental skyscrapers built in urban areas in the late nineteenth century.[40] New Almaden and New Idria were the quicksilver mines most fitting this description (Figure 3.13). In these mines, massive wooden members were arranged in a grid that held open the earth, allowing the miners space and relative safety in which to work. Similarly, in cities, steel members were joined in a spatial grid supporting tall buildings. As Brechin insightfully points out, skyscrapers used "much the same technology for vertical transportation, communication, and life support as did the Comstock Mines."[41] A few of the quicksilver mines in California and the West were upside-down cities, worked for many generations and eventually encompassing many levels with miles of tunnels and shafts.

The miners worked at creating space within the earth through a process of subtraction. The miner began in the open air and dug into the earth, creating the space necessary for his body to continue to dig. A miner might dig vertically, creating a shaft, or horizontally, creating a tunnel.[42] With shafts and tunnels as their primary tools, miners created the necessary space in the proper configuration to accomplish their work goal of extracting ore. These hollowed spaces within the earth were conduits serving human needs and desires, pathways for the flows of humans and air as well as the flows of waste rock and ore. Shafts and tunnels were the basic extensions of the surface world of humans into the earth that held their desire. Depending on the nature of the ore body, the subtracted space of the mine might be a tunnel adjacent to a narrow vein of metal that was large enough to allow the miner to work or a hollowed-out cavern—or *laboré*, as it was called at New Almaden—in which the miner would work in the space formerly filled with ore.

Aboveground landscapes of work at mercury mines included the reduction plant and the connected physical environments of the resources entering and waste leaving the plant (Figures 3.14 and 3.15). Reduction plants included spaces and structures for sorting and storing the ore, the furnaces and condensers for processing the ore, and ore dumps for the spent ore. In the process of developing a mine the largest initial capital outlay was for the construction of a reduction plant. Unlike the mine, which was a pay-as-you-go proposition, the plant had to be up and running early on for any mercury to be produced, although plants were often modified over time. Often a simple plant, such as retorts with pipe condensers, was constructed initially as a means of producing income while the extent of the ore body was probed. If the mine proved profitable, more elaborate furnaces might then be constructed.

As for the technology of reduction in the industry, it was adequate and largely based on practical experience, more so than on scientific experiment. The author of the *California State Mining Bureau Bulletin 78*, detailing the quicksilver resources of

FIGURE 3.13 Square Set Stope, New Idria Mine
This image shows square set stopes, between levels 2 and 2½ in the New Idria Mine. (Bradley, *Quicksilver Resources of California No. 78*, 111.)

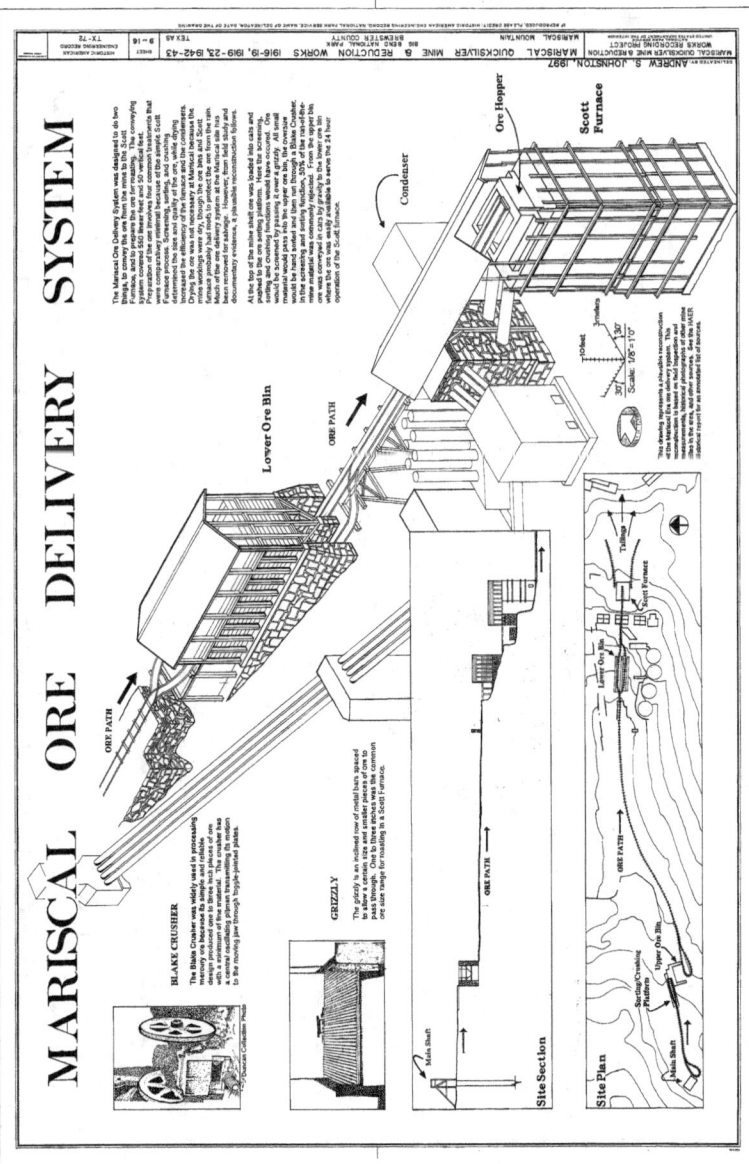

FIGURE 3.14 Mariscal Ore Delivery System

The Mariscal Mine ore delivery system, ca. 1920, in Big Bend National Park is detailed in this drawing. In this era the Mariscal Mine operated a four-shaft Scott furnace. (Library of Congress, Prints and Photographs Division, Historic American Engineering Record, National Park Service, Andrew Johnston, 1997.)

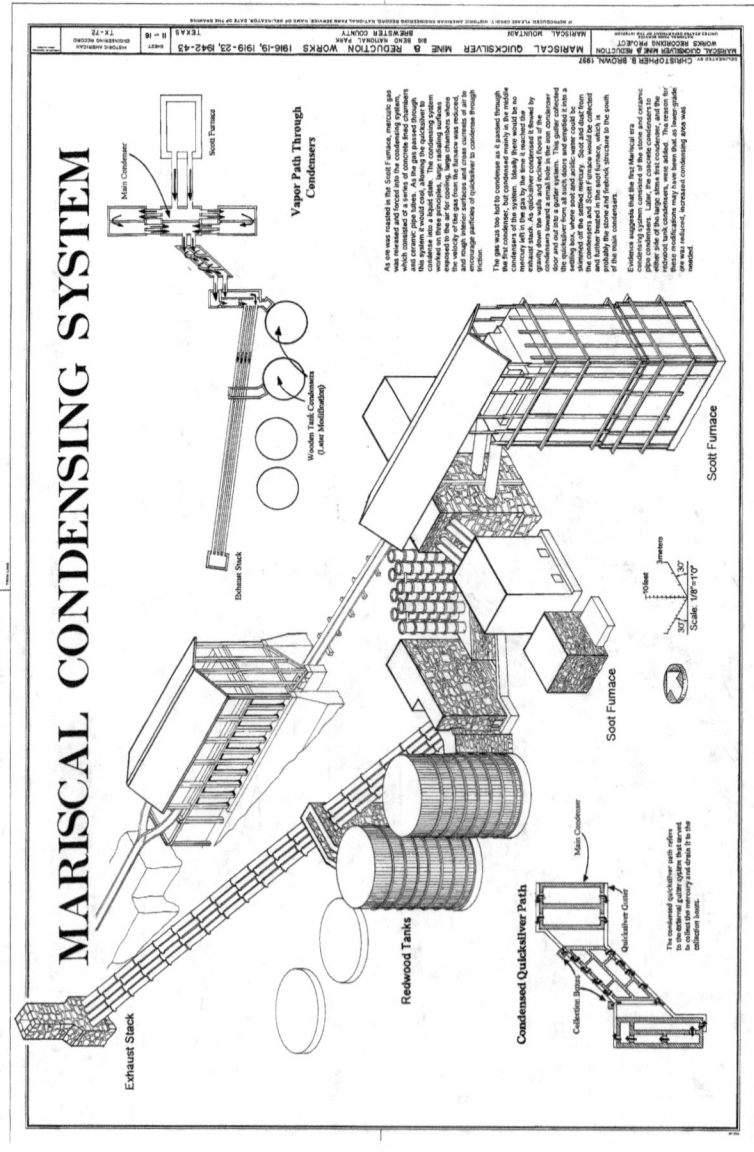

FIGURE 3.15 Mariscal Condensing System
The condensing system for the Mariscal Scott furnace included stone and concrete condensers, vitrified pipe condensers, and redwood tanks also used as condensers. (Library of Congress, Prints and Photographs Division, Historic American Engineering Record, National Park Service, Christopher B. Brown, 1997.)

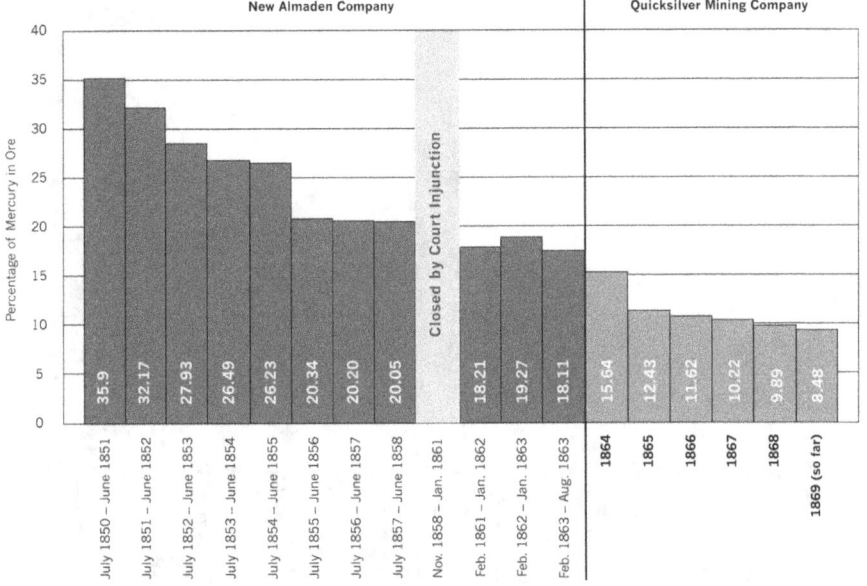

FIGURE 3.1B Richness of Ore Mined at New Almaden
Percentage of mercury in the ore mined at the New Almaden Mine, from July 1850 to 1869. Note the strong drop in richness over time. (New Almaden Mine Collection, 1845–1973, folder 7, Janin to Butterworth, January 12, 1869. Chart drawn by Austin Porter.)

California, stated, "Production of the metal is retarded, it is believed, by ignorance of its technology."[43] The author continued: "Experiments of various operators have been conducted, absolutely unaccompanied by any thought or attempt at technical control."[44] The reduction technology of the industry did undergo a major change during the quicksilver boom of the 1870s, when the economic and political factors behind the quicksilver price boom of the early 1870s were complicated by a change in the nature of the cinnabar ore coming from the mines. New Almaden felt this change acutely. The richest ore, coming from the mine in large chunks eight inches or more in diameter and with up to 30 percent mercury, had mostly been mined by the 1870s.[45] What remained were vast quantities of lower-grade, 3–8 percent ore. The mercury was the same whether it came from rich ore or poor ore, but new technology was necessary to process much greater quantities of the lower-grade ore in order to produce the same quantities of mercury. The changes made in reduction technology resulted in the ability to process more ore but did not necessarily result in greater processing efficiency as a comparison of production with cost.

A GEOGRAPHY OF MERCURY MINING IN CALIFORNIA 121

FIGURE 3.17 An Early, "Bustamante"-Type Furnace for Mercury Ore

This early furnace, developed at the Almadén Mine in Spain centuries before, was used at New Almaden early on. (A) is the firebox, (B) is the stacked chunks of ore, (C) marks the condensing chambers, (D) is the condensing trough, and (E) is the stack. This furnace is, in essence, the furnace drawn and pictured in Figures 1.1, 1.2, and 1.9. (St. John, "The Quicksilver Mine of New Almaden.")

When the Quicksilver Mining Company took over the New Almaden Mine in 1863, the mine was in extremely rich and plentiful ore (Figure 3.16). In the first year of the company's ownership (1864), it set a production record of 42,489 flasks, only to be topped in 1865 when it set the single-year record for the mine at 47,194 flasks. In this orgy of production, the company rapidly burned through the mine's richest ore and by 1874 the mine produced just over 9,000 flasks due to the lower-grade ore.[46] The rich ore that had been the mainstay of the mine in the 1850s and 1860s had been processed in intermittent "Bustamante"-type furnaces, a technology tailored to processing large chunks of ore (Figure 3.17). These furnaces, based on a centuries-old design imported from the Idrija Mine and modified for use in California, required men to carefully stack the chunks of ore in the furnace in such a way that the heat

FIGURE 3.18 Charging the Furnace
This image depicts men loading rich chunks of ore into an intermittent furnace at New Almaden, probably in the late 1850s or early 1860s. The men are working on a platform above the furnace, and handing ore down to men inside the furnace. (This image, like others in the J. Ross Browne article, is drawn from an extant photograph. J. Ross Browne, "Down in the Cinnabar Mines," 558.)

from the furnace could circulate freely through the ore (Figure 3.18). The alternative, if one had only fine ore, was to mix the fine ore with soil and water and place this mix into wooden forms to make sundried ore adobes. These adobes, roughly 9 by 4 by 4 inches in size, could then be stacked in the furnaces. However, making adobes added a significant cost to the reduction process, and the fine ore also tended to be of poorer quality.

When the Redington Mine challenged New Almaden and New Idria in the 1860s, it did so by processing large quantities of low-grade ore, much of which was fine ore that had to be made into adobes. The Redington Mine used the Idrija furnace, the most advanced quicksilver furnace in operation in California through the 1860s. For optimum operation, this furnace needed ores that were at least 5 percent mercury, forcing the Knoxville District mines not only to make adobes but also to make adobes out of ore that had been carefully sorted to reach a 5 percent concentration of mercury. The Idria furnaces were huge and expensive to construct; the ore chamber was 13 by 10 by 20 feet high, and the condensing chambers were even more massive (Figure 3.19). It took ten men one day to carefully stack 5,000 adobes in this furnace. Then, if fired on a Monday evening, the furnace was kept burning through Thursday

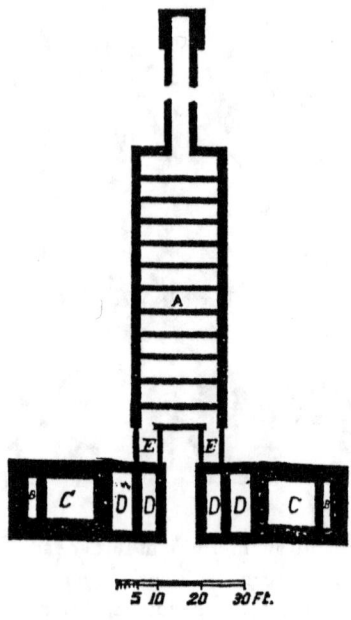

FIGURE 3.19 Plan and Section through an Old Idrija-type Furnace at the Redington Mine, Napa County, late 1860s to early 1870s

Egleston described the construction and operation of this furnace at length. Highlights of this furnace include its very great size (note the scale on the top drawing) and correspondingly its very great cost. In the top drawing (B) is the firebox; (C) is where the adobes were stacked; and (D), (E), and (A) are the condensers in series. Fumes would pass through holes from chamber to chamber in (A), giving time for the fumes to cool and the mercury to condense and accumulate in the bottom of the chambers. This was a double furnace with a shared condenser system (A), a design that fit the intermittent and staggered mode of operation. (Egleston, "Notes on the Treatment of Mercury in North California" 1875, fig. 2.)

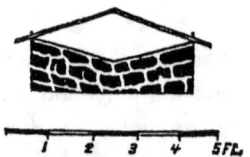

morning. The furnace was then allowed to cool for two days, and on Saturday was unloaded by six men and made ready for a new cycle the next week.[47]

The Redington Idrija furnaces were larger than the intermittent furnaces used at New Almaden, as an innovation meant to deal more efficiently with low-grade ore. And although the massive Idrija furnaces allowed the Redington Mine to process adobes more efficiently than at any other mine, they could barely compete with the very rich *grueso* ore and Idrija furnaces at the New Almaden Mine. Technologically what they needed was a way to efficiently process massive quantities of low-grade ore. The answer came in the form of an innovative furnace design developed in the nearby Manhattan Mine, also in the Knoxville District. This mine had low-grade ore like the Redington, which prompted the owners of the mine—Richard Knox and Joseph Osborn—to experiment with reduction technology in order to better process

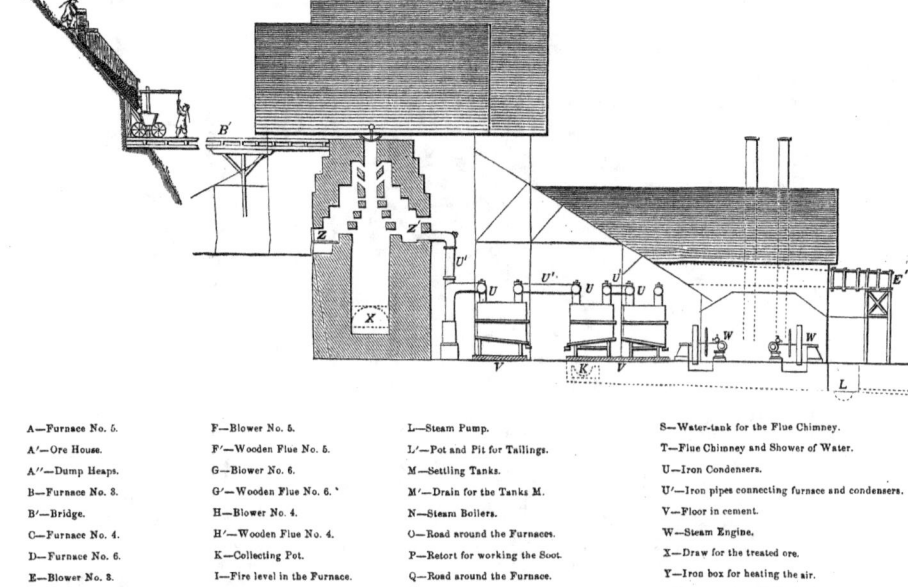

A—Furnace No. 5.	F—Blower No. 5.	L—Steam Pump.	S—Water-tank for the Flue Chimney.
A'—Ore House.	F'—Wooden Flue No. 5.	L'—Pot and Pit for Tailings.	T—Flue Chimney and Shower of Water.
A''—Dump Heaps.	G—Blower No. 6.	M—Settling Tanks.	U—Iron Condensers.
B—Furnace No. 3.	G'—Wooden Flue No. 6.	M'—Drain for the Tanks M.	U'—Iron pipes connecting furnace and condensers.
B'—Bridge.	H—Blower No. 4.	N—Steam Boilers.	V—Floor in cement.
C—Furnace No. 4.	H'—Wooden Flue No. 4.	O—Road around the Furnaces.	W—Steam Engine.
D—Furnace No. 6.	K—Collecting Pot.	P—Retort for working the Soot.	X—Draw for the treated ore.
E—Blower No. 3.	I—Fire level in the Furnace.	Q—Road around the Furnace.	Y—Iron box for heating the air.
E'—Wooden Flue No. 3.	J—Bottling Pit.	R—Water-tank for the main Condensers.	The arrows show the flow of the mercury.

FIGURE 3.20 Section through the Knox-Osborn Continuous Coarse-Ore Furnace at the Redington Mine, Napa County, 1874

This is a section through one of a series of Knox-Osborn furnace and condenser systems installed at the Redington Mine during the quicksilver boom years. Cinnabar ore entered the system from the top left, and mercury was collected in a trough at bottom right (L). The ore was fed into the furnace (the large brick structure near center), (X) was the draw for spent ore, (Z) denotes the firebox, and (Z') denotes the vapor chamber. The hot mercuric gas was sucked into the condenser system (U) by blowers (W). (K) was part of the mercury collection system. Note that the delineator drew Chinese men delivering ore to the furnace, although by all accounts the Redington Mine did not employ Chinese workers. Although the furnace depicted is a coarse-ore furnace, there was also a Knox-Osborn fine-ore furnace. (Egleston, "Notes on the Treatment of Mercury in North California," 1875, fig. 1.)

the low-grade fine ore coming from their mine. Knox and Osborn had realized this need years earlier and worked to invent a continuous-feed furnace that, unlike the intermittent Idrija furnaces, would operate constantly so that ore could be poured through it continuously. In January 1874 Knox and Osborn successfully operated a continuous-feed furnace that they claimed could process 900 to 1,000 tons of ore a week, as opposed to the 200 tons a week processed by the large Redington Idrija furnaces (Figure 3.20). What's more, rather than having to meet the 5 percent ore requirement, Knox and Osborn claimed that they could profitably process ore as low

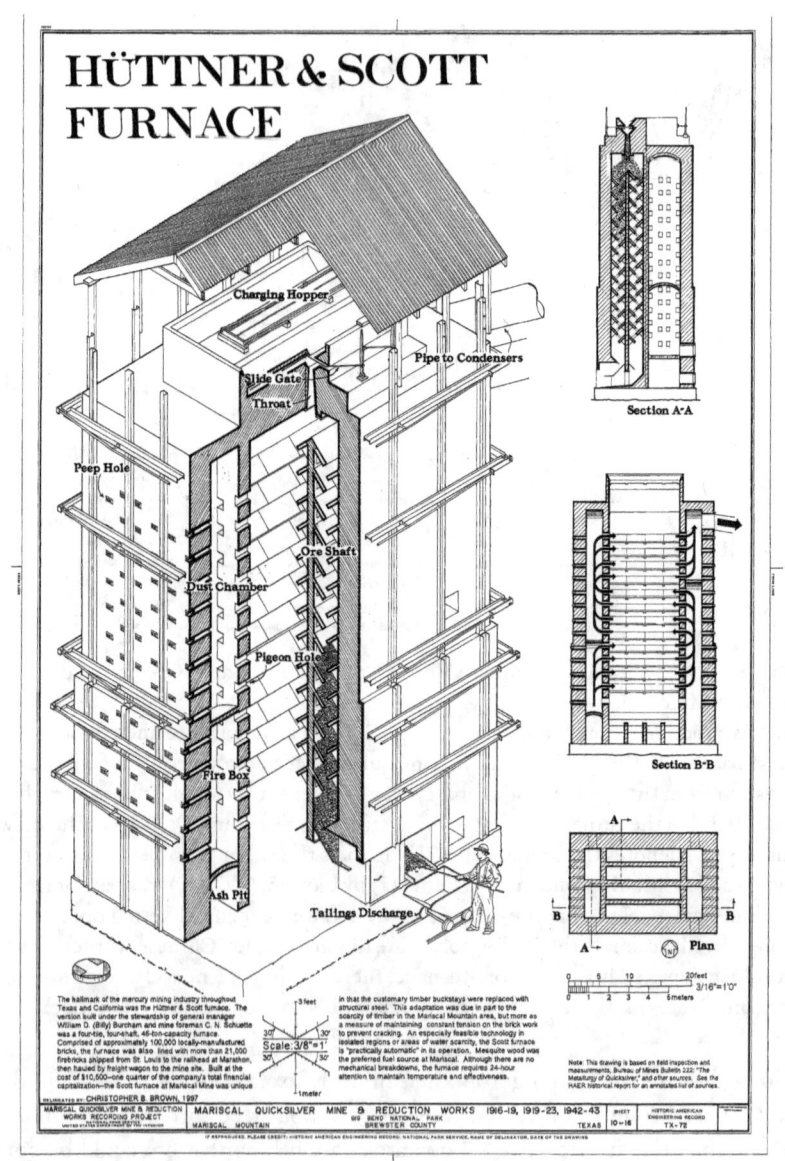

FIGURE 3.21 Hüttner & Scott Furnace

The Hüttner & Scott Furnace, commonly called the Scott furnace, was developed at New Almaden and became the most commonly used continuous fine-ore furnace design in the industry. (Library of Congress, Prints and Photographs Division, Historic American Engineering Record, National Park Service, Christopher B. Brown, 1997.)

as 0.5 percent.[48] The first Knox-Osborn furnaces were coarse-ore furnaces, requiring the ore to be in small chunks while processing no dust. Soon, however, Knox and Osborn invented a fine-ore continuous furnace that revolutionized the industry, allowing mines to crush cinnabar ore to the size of one to two inches in diameter, a much easier industrial process than making adobes. Also, cinnabar dust could be run through this furnace. Men at the New Almaden Mine, just a year behind Knox and Osborn, invented their own continuous-feed fine-ore furnace, called the Hüttner & Scott Furnace after the engineer and brick mason who developed the furnace (Figure 3.21).[49] With this furnace, by 1876 New Almaden produced over 20,000 flasks (up from the 9,000 flasks of 1874), processing large quantities of low-grade ore.

This new furnace technology was an equalizing force in the industry, giving any mine that could afford one the ability to reduce ore with nearly the same efficiency as the New Almaden Mine. The Knox-Osborn furnace was rapidly adopted by more than a dozen mines during the quicksilver boom. Factors such as the accessibility of ore, labor relations at the mines, transportation to and from the mines, and the management of mine camps determined the financial success of a given mine, whereas the ability to reduce ore was similar at all the mines. The new reduction technologies invented in California in the 1870s made differences in ore richness much less significant than they were with the Idrija furnaces used in the first decades of the industry.

As the production landscapes of quicksilver mines developed and changed, so did the quicksilver mining camps. Unlike rural development—in which a farmstead was joined by other farmsteads, which then over time supported the development of a farm-service town—mining camp development demanded support services (room, board, entertainment, etc.) to be available from the start, often for large numbers of workers. Mining camps could grow or shrink over time, but they always had to provide the basic services needed to support the workers. Early in the quicksilver industry, mine camps were most often company towns owned and operated by the mine. Later, during the quicksilver boom, the new mines contracted out the provision of these services to specialized companies. Quicksilver camps were always, however, structured by race, ethnicity, gender, and class. For instance, not only did mine companies hire groups of different ethnicities to do different tasks; they also felt they could provide for them and their families differently. The role of women in camp landscapes differed greatly over time and among mine types. Family life, for example, was the norm at New Almaden, whereas there were no women or families in the Chinese camps of the new mines.

During the first twenty years of the quicksilver industry the four large mines—New Almaden, Guadalupe, New Idria, and Redington—each developed substantial settlements, each a significant town in early California. With the quicksilver boom of the mid-1870s, there were suddenly scores of new quicksilver camps, all far less developed and substantial than those of the old mines. Many of the new camps were near existing towns, and whereas the old mines were relatively self-sufficient, the new mines relied heavily on the nearby towns for goods and services, often contracting out the provision of housing and food to local companies in these towns.

With the quicksilver boom, new mines and camps were developed throughout the Coast Ranges, many adjacent to existing towns. Residents of these existing towns viewed the new mines with some ambivalence. Local newspapers, for example, reacted with both pride and dismay. They recorded the weekly shipments from the mines and praised production increases, often describing the arrival of supplies and new machinery in detail. The society columns noted the comings and goings of owners and mine superintendents, while gatherings and parties at the mines were significant social events. The employment of Chinese workers at these new mines in the 1870s, however, provoked disapproval; as the *Independent Calistogan* reported on February 25, 1880, "The Great Western Quicksilver Mine, as conducted in the past, has benefited Calistoga, but not very materially. Had only white men been employed there would have been a great difference."[50]

The owners and managers of the new mines, of course, attempted to keep as much money circulating within their mining camps as possible, using the camp services as a way to recapture some of the money paid to employees just as the original mines had. From their point of view, a worker's pay was ideally paid back to the mine company for room and board and through purchases at the company store. For merchants in nearby towns, the company stores took away much of the potential retail trade with the workers. The mine company instead bought in bulk at wholesale prices from sellers both locally and in San Francisco. At the Great Western Mine—a new mine of the 1870s—the manager, Andrew Rocca, owned the company store with a Chinese merchant. The company store carried goods for both the 50 white workers and the 250 Chinese workers. Little of the Chinese workers' money found its way into nearby Middletown in part because of rules requiring the Chinese to purchase their goods at the company store. The white families, however, had less pressure to make their purchase at the company store and probably contributed substantially more to the town economy.

A major source of economic benefit to the towns came from selling supplies to the mines. The supplies needed included timber, tools, building materials, and machines and equipment. Although some of these supplies were available locally, others—such as machines and equipment—came in on the train and were then shipped by teamsters to the mine. These same teamsters also transported the heavy cast-iron flasks of quicksilver from the mines to the towns' train depots (Figure 3.22). At the train depots, the number of flasks was recorded and published in the town paper.

FIGURE 3.22 Flasks for Shipment
A shipment of 300 flasks of quicksilver from the New Idria Mine. (Bradley, *Quicksilver Resources of California No. 78*, 10.)

Records for the Guadalupe Mine for November and December 1880 show that of total mine expenditures, 70–75 percent went to labor. Of this, the mine recaptured what it could through room and board and purchases at the company stores. The remainder of the mine expenditures, 25–30 percent, went to material costs and sundries. During these two months about half (48%) of the sundry expenditures were paid to San Francisco firms, 40 percent went to San Jose firms (this included all the wood and timber charges), and 10 percent went to local firms and individuals located at or very near the mine.[51] As these records show, the regional and local trade of the quicksilver mines contributed to the development of many Coast Range communities, and the regional center of San Francisco.

Four Competing Models

By the 1870s there were four competing models of quicksilver mining in California, each producing mercury for the market. Each model was a social and cultural creation. The distinction between the original mines and the new mines of the 1870s is central to the story of the industry. The quicksilver booms of the mid-1870s caused a fundamental shift in the industry by fostering the development of the new mines, transforming mercury's role in the development of California. Race and ethnicity were the defining characteristic of each of these models, and the landscapes of work and camp life in each model were markedly different as a result.

The next three chapters explore the landscapes of work and the landscapes of camp life within the mercury mining industry in great detail. This chapter has set the stage by arguing that the geography of the mercury mining industry is not a simple mapping of the geology of cinnabar, but that geology intersects with social factors such as money and power, and technology and function, to create the geography of the mercury industry in California. The organization of quicksilver mines and camps tells us about the development of California, the same way that the mines and camps of the gold and silver industry have been used to do this. However, the history of quicksilver has a different story to tell.

Notes

1. Kenneth Rank–New Idria Mines Collection, 1867–1973, box 1559, folder 6, "E. R. Sampson to the President and Directors of the Quicksilver Mining Co., New York, July 7, 1869."

2. The New Almaden Mine ceased operation due to court injunction on October 30, 1858. Ownership of the New Idria Mine in the nineteenth century has not been fully documented. William Barron, Thomas Bell, and other partners purchased control of the mine in 1858. These other partners at this time, 1858, are not known. However, in an 1866 newspaper article announcing a stockholders' meeting, Wm. H. L. Barnes, W. C. Ralston, D. O. Mills, and Wm. Neely Thompson are named as the trustees of the New Idria Mining Company, clearly giving control of the mine to the Bank Crowd. *San Francisco Bulletin*, published as the *Evening*

Bulletin 21, no. 76 (January 6, 1866): 4. An 1870 article lists W. E. Barron (president), Thomas Bell, D. O. Mills, W. C. Ralston, and William Burling as trustees of the New Idria Mine. *San Francisco Bulletin*, published as the *Evening Bulletin* 30, no. 16 (April 26, 1870): 3.

3. Kenneth Rank–New Idria Mines Collection, 1867–1973, box 1559, folder 6, "E. R. Sampson to the President and Directors of the Quicksilver Mining Co., New York, July 7, 1869." Additional information on the New Idria Mine comes from Walter W. Bradley, *Quicksilver Resources of California*, 93–120.

4. Kenneth Rank–New Idria Mines Collection, 1867–1973, box 1559, folder 6, "E. R. Sampson to the President and Directors of the Quicksilver Mining Co., New York, July 7, 1869."

5. Ibid.

6. A great deal of the success of the New Idria Mine in the twentieth century was due to the rotary furnace developed there about 1918. This furnace design proved to be very efficient at processing large quantities of low-grade ore and was the industry standard up until the end of mercury mining in the United States in the 1970s.

7. Kenneth Rank–New Idria Mines Collection, 1867–1973, box 1559, folder 6, "E. R. Sampson to the President and Directors of the Quicksilver Mining Co., New York, July 7, 1869."

8. For information on the early connections among Monterey, San Juan, and New Idria, see Thomas Oliver Larkin et al., *The Larkin Papers*, Josiah Belden to Larkin, vol. 5: 361.

9. Irving Murray Scott Correspondence, November 24, 1871–March 1, 1882, BANC MSS 2009/138.

10. See also Johnston, "Quicksilver Landscapes."

11. For more on multiple and conflicting understandings of geology in the nineteenth century, see Mott Greene, *Geology in the Nineteenth Century*.

12. George F. Becker, *Geology of the Quicksilver Deposits of the Pacific Slope*.

13. Ibid., 417–18.

14. Curt N. Schuette (1895–1975) was a mining engineer and geologist who specialized in mercury, working in the industry from before World War I until his death. Schuette was one of the twentieth century's most knowledgeable experts on quicksilver mines and mining in the United States. One of his first jobs was running the Mariscal Mine in the Big Bend of Texas during World War I, and during World War II he established a rotary kiln reduction plant, the ruins of which still exist on Mine Hill at New Almaden. The article discussed here is Curt N. Schuette, "The Geology of Quicksilver Ore Deposits." Years earlier J. A. Udden, director of the Bureau of Economic Geology and Technology, University of Texas, presented similar theories on the formation of quicksilver deposits based on his research in Texas. J. A. Udden, "The Anticlinal Theory as Applied to Some Quicksilver Deposits."

15. Samuel Benedict Christy, "On the Genesis of Cinnabar Deposits." Christy argued against a theory that cinnabar deposits were the result of sublimation, or the uplifting of deposits formed deep in the earth.

16. Schuette defined three quicksilver mining provinces in the United States: the California province, the Nevada or Basin Range province, and the Texas province. In his article, he

presented a generalized geologic picture for the occurrence of mercury in each of these provinces. Schuette, "The Geology of Quicksilver Ore Deposits," 47–50.

17. Quicksilver mining in the Big Bend mostly occurred in the early twentieth century. The most significant mine in the area was the Chisos Mine. See Kenneth B. Ragsdale, *Quicksilver*.

18. According to modern geologists, the Coast Ranges were formed by the action of the Pacific plate as it dove under the North American plate along the San Andreas Fault and other, minor faults. Rocks that were on the ocean floor were pushed up by the action of these plates and formed the mass of the Coast Ranges. The Coast Ranges consist of two major groups of rocks: the Franciscan complex and the Great Valley sequence. The Franciscan complex is a mélange, or chaotic jumble of sedimentary rock formed on the ocean floor at different depths and different times, combined with basalt from the ocean floor. This rock was subsumed into ocean trenches, mixed, and then forced up to form the Coast Ranges through the action of the plates. Often in the Coast Ranges, due to the chaotic jumble of the rocks, one rock outcrop is very different from the next, making geologic detective work, including finding cinnabar deposits, exceedingly difficult if not impossible. The most abundant rock of the Franciscan complex is muddy, low-density sandstone, and it is in this porous sandstone that many cinnabar deposits were found. The Great Valley sequence is a layer of ocean sediments that overlays the Coast Range ophiolite (a former slab of ocean floor with serpentine at its base) and is generally intact despite being tilted. Generally this rock is sandstone between layers of black shale. Cinnabar was also deposited in the sandstone of the Great Valley sequence. See Arthur D. Howard, *Geologic History of Middle California*; and David Alt and Donald W. Hyndman, *Roadside Geology of Northern and Central California*.

19. New Almaden was an exception, with working depths nearing 1,800 feet. However, the richest ore at New Almaden was found above 600 feet.

20. A number of geologists claimed active deposition at the Sulphur Bank Mine on Clear Lake. See Becker, *Geology of the Quicksilver Deposits of the Pacific Slope*.

21. As quoted in Jesse D. Mason, *History of Santa Barbara and Ventura Counties*, 185.

22. Walter W. Bradley, ed., *Quicksilver Resources of California*.

23. An example of the wheeling and dealing going on for mines during the quicksilver rush is the following from the *Mining and Scientific Press*, April 4, 1874, 213, regarding the Pine Flat District. "Sonoma County. Quicksilver Mining items—The Lost Ledge Mine has been leased to John A. Robertson and Co., of San Jose, for one or two years, with the privilege of buying the mine during the lease at $40,000. The Kentuck has been leased to W. A. Stewart, for $300 per month, during one year, with privilege of purchase during the time, for $30,000. The Rattlesnake having been sold and incorporated, a considerable force has been set to work. We learn from the best authority that the sale was without reservation for $52,000 cash. McKay, Snow & Sleeper, former owners, are still at Pine Flat."

24. Tiburcio Parrott leased the Great Eastern Mine in Sonoma County in the mid- and late 1870s, which proved to be profitable for both parties. The mine was a proven producer, however. Most prospects were sold outright, and the sellers at times reaped healthy payoffs. The

purchasers of the Great Western Prospect were often named as the most successful purchasers of a quicksilver prospect during the boom years.

25. A London newspaper clipping in a scrapbook of J. B. Randol's shows that the owners of the Redington Mine tried to market their mine to British investors, offering 22,000 shares of the mine for ten pounds each. The advertisement, undated but listing dividends from 1873, stated that this was the first quicksilver mine ever placed upon the British market. Randol, *Papers Relating to Quicksilver Mining, Ca. 1849–1894*, vol. 3. For more on British investment in American mining, see Clark C. Spence, *British Investments and the American Mining Frontier, 1860–1901*.

26. Henry Pitts, "Proceedings of the Miners' Convention in the San Carlos District in the Counties of Monterey and Mariposa and State of California," in Kenneth Rank–New Idria Mines Collection, 1867–1973. Kenneth Rank notes on the Pitts document that it is from the San Benito County Recorder's Office, and dated 1854. For the purposes here, I take this document at face value. However, given the long and corrupt McGarrahan legal proceedings regarding the mine, any document on New Idria must be very carefully researched and understood regarding its authorship and meaning.

27. Shinn, *Mining Camps: A Study in American Frontier Government*, 117–25.

28. Henry Pitts, "Proceedings of the Miners' Convention in the San Carlos District in the Counties of Monterey and Mariposa and State of California," in Kenneth Rank–New Idria Mines Collection, 1867–1973.

29. The ultimately unsuccessful Josephine Mine in Southern California was another prospect purchased and invested in by Barron & Co. See the listing for this mine in Lewis E. Aubury, *The Quicksilver Resources of California*.

30. Joseph Daniel Pelanconi, "Quicksilver Rush of Sonoma County, 1873–1875," 2–4.

31. Fortunately for the historian studying mining, there are often copious records of the goals, means, and methods of mining left by the economic geologist. Often employed by federal or state governments as well as by universities, economic geologists studied all mining activity in the United States for the purpose of promoting and developing mining as an industry. Their methods involved understanding the markets for mined materials, the geology of the individual mines, and the state of available mining and ore processing technology. Economic geologists then made informed analysis of the economic viability of particular mines, mining districts, and whole mining industries. A typical report on a mine forecasted the markets, examined the geology of the ore deposit, addressed the methods that the mining company employed to extract and process the ore, detailed their costs, suggested improvements, and often made projections on the future of the site.

32. Napa and Lake Counties both wanted the Redington District mines contributing to their tax base. Napa County eventually won this battle, and an odd wiggle in the county boundaries to this day is a lasting reminder of this battle.

33. Anthony Zellerbach financed the development of the Altoona Mine in northern California and the Oceanic Mine in San Luis Obispo County. Bradley, *Quicksilver Resources of California Bulletin No. 78*.

34. Bancroft Library, MSS C-F, 196, "Diary Relating to Mercury Mining." This diary/account book was written by Nathaniel Frothingham Disturnell (sometimes spelled Disturnall), 1840–1917, near Wilbur Springs, California, from 1874 to 1876.

35. This information comes from Nathaniel Disturnell's diary.

36. See "feature system" in Donald Hardesty, *The Archaeology of Mines and Mining*.

37. This idea of the nature of western mining development has been promoted by various authors to counter the long-held theories on frontier development argued by Frederick Jackson Turner, *The Frontier in American History*. These authors include Patricia Nelson Limerick, *The Legacy of Conquest*; and Paul and West, *Mining Frontiers of the Far West, 1848–1880*.

38. T. A. Rickard, *Four Mining Engineers at New Almaden*.

39. See chap. 1 in Brechin, *Imperial San Francisco*.

40. The full name for this method of timbering is the Deidesheimer square set.

41. Brechin, *Imperial San Francisco*, 69.

42. Miners also dug at all angles between the horizontal and the vertical. These features were generally called inclined tunnels or inclined shafts.

43. Bradley, *Quicksilver Resources*, 207.

44. Ibid., 208.

45. The New Almaden Mine established the names and standards for cinnabar quality that were then adopted throughout much of the industry. *Grueso* was high-quality ore in chunks eight inches or more in diameter. *Granza* was similarly rich ore but in chunks of three to eight inches. *Granzita* was generally poorer ore in even smaller pieces. *Tierra* was the name given to fine ore and ore dust and was also generally poorer in quality. Randol, *Papers Relating to Quicksilver Mining, Ca. 1849–1894*, vol. 3.

46. Schneider, *Quicksilver*, 49–76.

47. T. Egleston, "Notes on the Treatment of Mercury in North California." This description of the operation of an Idrija furnace is based on one operating at the Redington Mine in the early 1870s.

48. Ibid., 293.

49. Knox and Osborn sued most of the major quicksilver mines in the late 1870s for patent infringement on their continuous furnaces. The Great Western, Guadalupe, and other mines settled out of court. The New Almaden Mine—sued for the development of the Hüttner-Scott Furnace, which Knox and Osborn claimed infringed on the patents of their continuous ore furnaces—fought aggressively, sending operatives to Europe to buy rights to continuous furnaces there. Eventually New Almaden successfully argued that it did not infringe on the Knox-Osborn patents, and it made a deal with them to split the royalties on either furnace design. New Almaden Mine Collection, 1845–1973.

50. "The Leprous, Almond-Eyed, Heathen Chinese Must Go! One Corporation Shows 'Sand,' and Does Not Acquiesce," *Independent Calistogan*, February 25, 1880.

51. Guadalupe Mine Collection, 1854–1921, C-G 65, box 2, folder 2. Purchases from San Francisco included hardware, powder and fuse, oils, and miscellaneous equipment. Although the expenditures to San Jose were mostly for timber, there were also bills for horse feed and

wagon repair. Local expenditures included lime, ice, and candles. Only about 2 percent of expenditures were paid to the railroad and to a hauling contractor, a figure that is low even considering that the mine was very close to San Jose. The company store was run as a separate business.

4

Race, Space, and Power at New Almaden

The New Almaden Mine was created to exploit the rich cinnabar deposits in the hills south of San Francisco Bay. Like any remote industrial site constructed for resource exploitation, the mine was composed of work landscapes and camp landscapes, and these were created through the struggles of various groups of people involved with the mine, each attempting to further their own interests.[1] These groups of people can be defined by many factors, including race, ethnicity, class, and gender. Although these factors were intertwined, close study of the mines and camps of New Almaden and other quicksilver mines in the nineteenth century demonstrates the importance of the racialization process to the structuring of the work and camp landscapes of the industry, and the change of these landscapes over time.

Historians have made a strong case for the existence of a racial hierarchy in California in the nineteenth century and the defining role it had in the development of the state.[2] In *Racial Fault Lines: The Historical Origins of White Supremacy in California*, Tomas Almaguer argued that "race and the racialization process in California became the central organizing principle of group life during the state's formative period of development."[3] Although issues of race and class were intertwined, for Almaguer it was race, not class, that was the central organizing principle behind inequality, because in part race determined who had access to what class strata and opportunities. An ethnic group's power and influence in the state determined how that group was racialized and fit into hierarchies and, relatedly, their class possibilities. In California, a state with many races and ethnicities, the racial hierarchy was greatly contested, and as presented by Almaguer consisted of whites occupying the most privileged position, followed by these groups in order of decreasing power and influence: Californios, Mexicans and other Spanish-speakers, blacks, and Indians.

In the history of the quicksilver industry, racial and ethnic classifications continually rise to the top when exploring

DOI: 10.5876/9781607322436:c04

questions concerning the shaping of the landscapes of California's quicksilver mines and camps. Interestingly, the racial and ethnic groups comprising the hierarchy changed during the quicksilver boom years of the 1870s and early 1880s; in these years Chinese workers emerged as the most numerous single racial and ethnic group in the industry. As a group the Chinese came to occupy the racial hierarchy just above Indians. This kind of insertion into the racial hierarchy was typical—new groups were positioned in relation to other groups without otherwise disturbing the existing hierarchy. As will be detailed in this chapter and in chapters 5 and 6, during the quicksilver boom years there were many different types of mines, and many modes of mine and camp organization. However, despite these differences, the racial hierarchy at the mines always remained intact. Looking at the quicksilver industry through the lens of racial hierarchy leads to the conclusion that although the quicksilver industry, and the landscapes of its mines, underwent radical change from the 1840s to the 1890s, the racial and ethnic hierarchy of California and also of the quicksilver industry remained essentially the same over this time, with groups being added or removed based on their employment in the industry. Although jobs were done by different racial and ethnic groups from mine to mine, groups lower in the hierarchy always did lower-status jobs. In the quicksilver boom, there were many mines with a variety of different racial and ethnic workforces and a variety of types of work and camp landscapes, but the racial and ethnic hierarchy was never violated. The boom proved that there were many ways to create and operate mines to exploit cinnabar resources; however, the racial and ethnic hierarchies of California at the time dictated an inviolable structure to the industry.

This chapter explores the landscapes of work and camp life at New Almaden from its founding in the 1840s, through the transformations of the change in ownership in the 1860s, to the quicksilver boom and bust of the 1870s. This exploration focuses on both the contract system of mining imported by Barron, Forbes & Co. and how race and ethnicity were central features of this system, and the racial and ethnic structuring of the various camps and camp life at New Almaden. Chapters 5 and 6 then look at questions of race, ethnicity and landscape at the other quicksilver mines of California during the boom and bust years of the 1870s and beyond, exploring the competing mining models and what the experience of the quicksilver industry can teach us about the development of California.

Creating a Racialized Workforce at New Almaden

Workers at the quicksilver mines had to perform two industrial processes: they had to mine the ore, and then they had to process the ore to produce mercury. Mercury mining was "hardrock" mining, involving digging tunnels and shafts through rock to reach and then exploit bodies of ore. Once this ore was brought to the surface it was processed by reduction, that is, heating the ore until the mercury content vaporized, and then cooling the resulting mercuric gas until it condensed into liquid mercury. There were many ways for mine managers and workers to accomplish these two basic functions of mining and reduction.

From about 1849 to 1869 the New Almaden Mine dominated the mercury industry in California and, as detailed in chapter 1, Barron, Forbes & Co. developed the mine as they had many of their other ventures, using an amalgam of British and Spanish colonial models for developing businesses on the frontier. Whereas before the gold rush the mine was worked by Indians overseen by Californios and Mexicans, by 1850 New Almaden was worked primarily by Mexicans and Chileans recruited by Barron, Forbes & Co.[4] These recruits were experienced hardrock miners coming from long-established silver mining districts in Mexico and Chile. One such miner was Juan Bañales (d. 1871), whose grave at New Almaden says he was born in 1809 in the village of San Sebastián, Sinaloa, Mexico. This hill town is about thirty miles inland from Mazatlán, a major port for Barron, Forbes & Co.

By the mid- and late 1850s, Barron, Forbes & Co. modified its original reliance on Mexican and Chilean workers by hiring large numbers of Cornish miners complemented by a smattering of miners from other European countries and from America.[5] These two groups of workers practiced very different methods of mining. The original Mexican and Chilean workers at New Almaden practiced a method of mining called El Sistema del Rato—"the system of the moment"[6] (Figure 4.1). Derisively called the "rat-hole" method by American mining engineers dismayed at the irregular passages leading in all directions, the system was better described as the practical or empirical system. Miners working with this method followed bodies of ore wherever they led, digging out the ore and only as much of the surrounding rock as was necessary to get the ore that was in sight (Figure 4.2).[7] New Almaden was first developed with this system, and an early visitor to the mine, Mrs. S. A. Downer, said in 1854 of the resulting underground landscape: "The crossings and re-crossings, the windings and intricacies of the labyrinthine passages could only be compared to the streets of a dense city, while nothing short of the clue, furnished Theseus by Ariadne, would insure the safe return into day, of the unfortunate pilgrim, who should enter without a guide."[8] El Sistema del Rato was imported to the New World by the Spanish in the sixteenth century, and was a well-developed, long-practiced system that was very efficient in removing only valuable ore at a time when both ore and waste rock had to be packed out of mines on the backs of men (Figure 4.3).

The Cornish and European miners hired at New Almaden were trained in "advanced" mining methods developed in conjunction with systematic mine surveying, advanced

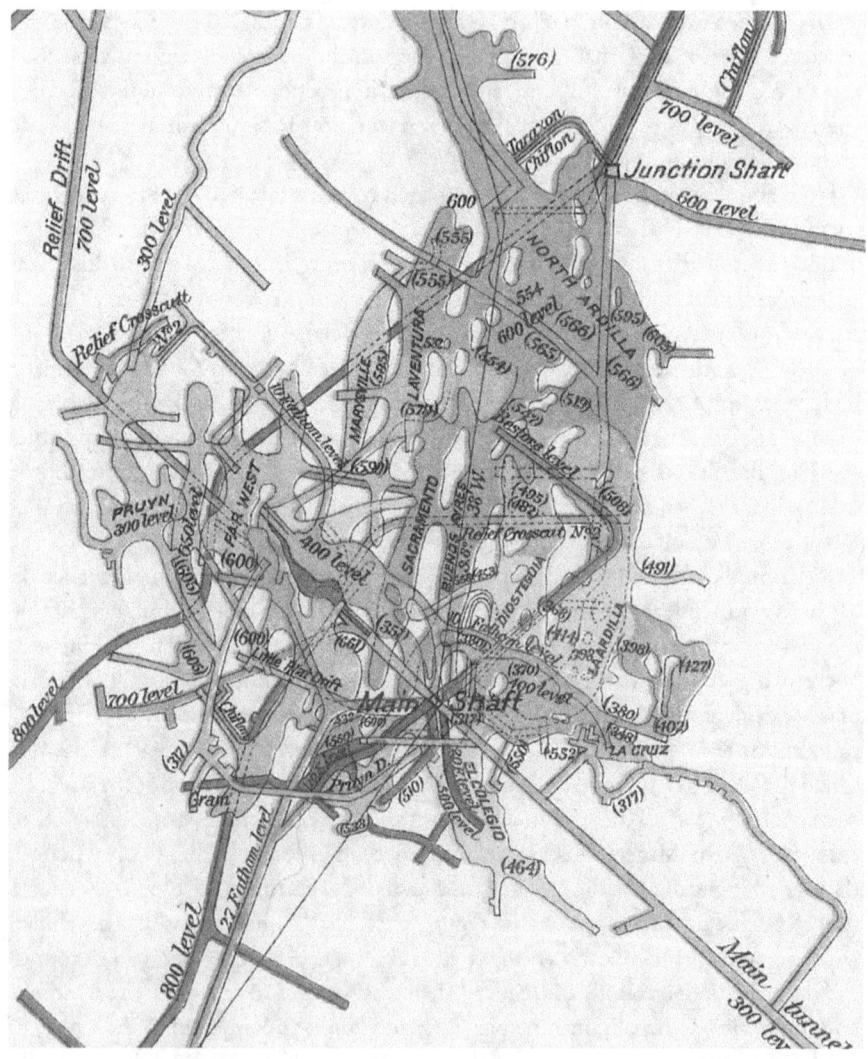

FIGURE 4.1 Plan of Mine Workings at New Almaden

El Sistema del Rato, or "the system of the moment," was a mining method used by Mexican miners in the earliest days of work at New Almaden in which miners followed ore wherever it led, removing only that rock necessary to follow the ore. The relatively early workings depicted in this plan, while not an example of El Sistema del Rato, nonetheless display some of its characteristics, and were probably mined by miners used to working in this mode. True examples of El Sistema del Rato from its earliest days were erased in later mining efforts at New Almaden. (Detail of a plan compiled by F. Reade to accompany Becker, *Geology of the Quicksilver Deposits of the Pacific Slope.*)

FIGURE 4.2 Francisco Velásquez Chamber
In the early years of mining at New Almaden, Mexican miners excavated great chambers of very rich ore, leaving large, cavernlike spaces that when excavated showed the original shape of the cinnabar body. Leaving pillars of ore as supports is sometimes referred to as the "Mexican Method." (Browne, "Down in the Cinnabar Mines," 556.)

mathematics, and understandings of geologic theory. Rather than empirically following a vein of ore as miners did in El Sistema del Rato, the miners of the advanced system constructed elaborate systems of mine infrastructure based on theories by geologists of where ore was to be found and on the mathematics of mine surveyors and engineers about how to get there. The resulting underground landscape of these advanced techniques is well illustrated by the plan and section of the New Almaden Mine from the early 1880s (Figures 4.4 and 4.5). The section shows shafts descending over 1,500 feet into the earth from which tunnels ran horizontally at regular intervals, and the plan shows how tunnels radiated from these shafts.[9] Mine development at this scale involved removing vast amounts of waste rock in order to get to the ore,

FIGURE 4.3 *Tenateros* (Tenateros) Carrying the Ore from the New Almaden Mine, early 1850s

El Sistema del Rato was based on making expeditious use of the spaces created in the mines. Although exaggerating the space in the mines, this etching does show the difficult climbs and descents that ore and rock carriers had to negotiate during the early days of working the mine. (James M. Hutchings, *Scenes of Wonder and Curiosity in California.*)

FIGURE 4.4 Plan of the Underground Workings at the New Almaden Mine, early 1880s
This detail shows the underground workings in the vicinity of the Randol shaft, located in the center-right of the image. Yardage contractors made the shafts and tunnels, "tributors" dug out the ore bodies, and "trammers" took the ore from the ore bodies to the shafts. It is useful to compare this image (especially the Randol shaft, center right) with the section of the Randol shaft in Figure 4.5. Ore bodies descend from the bottom right to the top left in this plan. (The section in Figure 4.5 shows the ore deposits on the left side of this image.) Once this general trend of the ore was determined, mine managers sought to build the mine infrastructure to go deeper, and this involved deepening the Randol shaft and developing the radiating tunnels. This plan shows the parts of the mine developed during the years 1875–83. (Plan compiled by F. Reade to accompany Becker, *Geology of the Quicksilver Deposits of the Pacific Slope*.)

and as a result required elaborate lifting mechanisms in the shafts. Furthermore, deep shafts and tunnels often required massive pumping operations to keep the mines from flooding. Technologies such as steam power were necessary for these "advanced" mining techniques (Figure 4.6).

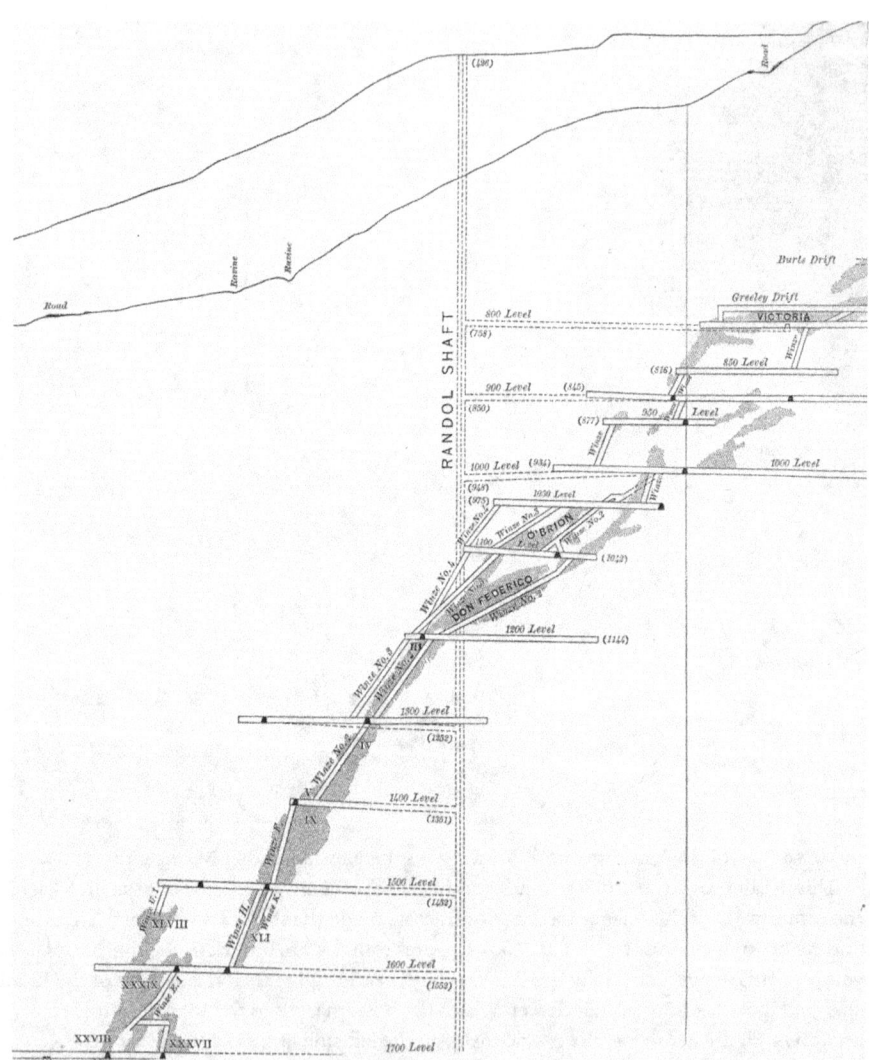

FIGURE 4.5 Section of the Underground Workings at the New Almaden Mine, early 1880s

This section shows the Randol shaft, tunnels radiating from the shaft at 100-foot intervals, and the ore bodies (shaded areas). The location of this section cut is shown in the plan in Figure 4.4—the cut is a thin line that makes a triangle across the bottom left corner to the top left corner of the plan. Matching the names (such as Don Federico) and numbers of the ore bodies from section to plan helps. The shift from naming to numbering ore bodies happened in 1877. The Randol shaft is actually in the background, with tunnels extending toward the descending ore body shown in the section above. (Section compiled by F. Reade to accompany Becker, *Geology of the Quicksilver Deposits of the Pacific Slope*.)

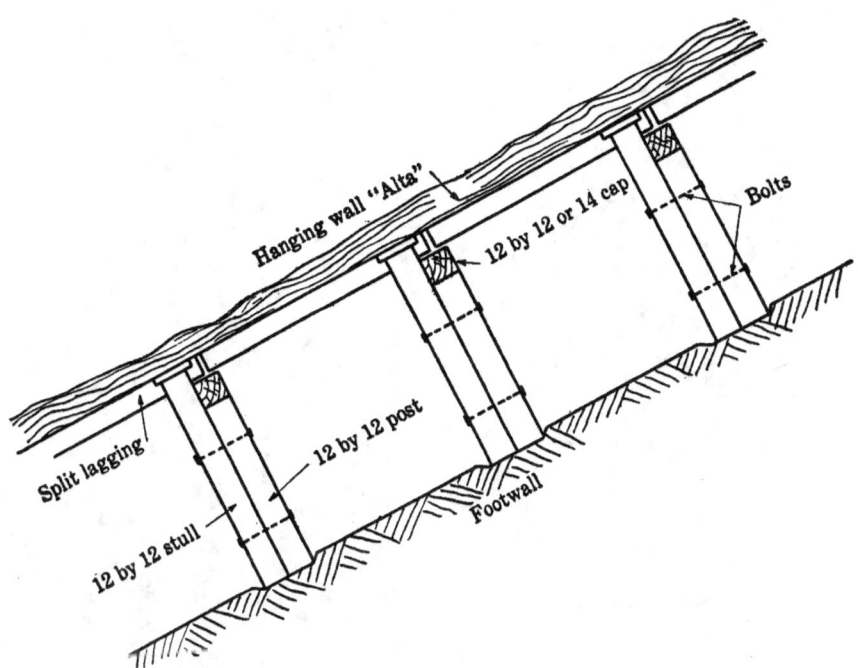

FIGURE 4.6 Stope Timbering

Once "advanced" mining techniques were instituted at New Almaden, the method used for extracting the ore was called overhand stoping. Once tunnels were opened above and below the ore body, the extraction began on the lower level with the miners working from the hanging wall to the footwall (left to right on the ore bodies as drawn in Figure 4.5; see terms in Figure 3.12). As miners worked upward in the ore body they placed "stulls," or supports, constructed of twelve- by twelve-inch timbers. This was stope timbering. These stulls supported the hanging wall and gave the miners a platform on which to work. (Schuette, "The Geology of Quicksilver Ore Deposits," 20.)

Work was organized at New Almaden according to a contract system whereby groups of miners bid on limited-term work contracts as specified by the company. Certain types of contracts most often went to particular racial and ethnic groups; specifically, Mexican miners worked tribute contracts in which they mined ore, and Cornish miners worked yardage contracts in which they extended the mine infrastructure. The New Almaden contract system was a hybrid of Spanish colonial mining organization familiar to the Mexican and Chilean miners and European, particularly Cornish, mining methods familiar to the European mine managers and Cornish miners (Figures 4.7 and 4.8). Miners in Mexico and Chile worked in a tribute system;

FIGURE 4.7 Patricio Avila, a Spanish-speaking miner, New Almaden, c. 1880 (Courtesy of History San Jose Research Library, Historic Images of the New Almaden Quicksilver Mines, 1979-251-138.)

FIGURE 4.8 William Doige, Cornish Miner, New Almaden, c. 1880 (Courtesy of History San Jose Research Library, Historic Images of the New Almaden Quicksilver Mines, 1979-251-124.)

that is, a mining guild of which they were a part worked a mine for a share of the profit. Miners in Cornwall worked in groups either on contract or by tribute; contract work was for the development of mine infrastructure, whereas tribute work was for extracting ore in the ore bodies.[10]

The New Almaden contract system was a hybrid recognizable to both traditions while allowing for multiple racial and ethnic groups at the same mine. Due to New Almaden's dominance in the industry, other mines such as New Idria and Guadalupe mimicked the contract system and created similarly racialized workforces. New Almaden had the distinction of being the largest mine of any type in the American West worked solely on contracts, and the original mines of the quicksilver industry were the only major mines in the West to use contract mining to organize most work. The reasons for this unique aspect of quicksilver mining are complex. Key factors are the Spanish colonial ties through Barron, Forbes & Co., the monopoly character of the industry, and the complex racial and ethnic structure of the workforce that was manipulated successfully by mine owners and managers in part to deny worker groups the ability to organize successfully against management. The contract system worked at New Almaden (while not at other large mines in other mining industries) because few of the miners at New Almaden came from a tradition of wage labor. Wage labor was common in American gold and silver mining, where miners organized to maintain their wages. Spanish and Cornish miners, however, were accustomed to being part of a mining company, and it was this group, not the individual, that dealt with the mine owners and managers. Individual miners were compensated for their work by the group according to their position and power within that group.

Few records exist of the New Almaden Mine under Barron, Forbes & Co., making it difficult to quantify the relative positions of Spanish versus English workers at the mine during the early years of operation. However, the Cornish miners probably had a higher status because they were the ones recruited for their skill in opening the large and deep deposits in the mine, making them valuable workers. They were white and Protestant as well, placing them high on the racial hierarchy. Later records for the mine (after 1865) show that Cornish miners were paid on average a third more than Spanish-speaking workers.[11] Following California statehood, Mexicans in California—as well as Spanish-speaking Californios—rapidly lost their status in California. Although the treaty of Guadalupe-Hidalgo that ended the Mexican-American War in 1848 established that people of Mexican descent in California were white by law and entitled to citizenship, in the 1850s there was a transition among the ever-growing Anglo population to the notion of the innate superiority of the American Anglo-Saxon branch of the Caucasian race.[12] Despite the treaty, after the war with Mexico, non-Hispanic whites increasingly enforced their status as superior and desired the property and resources that were held by Californios.[13]

Barron, Forbes & Co. knew from the mid-1850s that losing New Almaden (because its claim was based on a Mexican land grant) was a real possibility, and indeed, as we saw in chapter 2, the arguments against the Mexican land claim won out, with Barron, Forbes & Co. losing the mine in 1863. In the interim, its goal became to produce as much mercury as quickly as possible to flood world markets and to create a stockpile. Creating a racialized mine workforce suited this goal, allowing it to develop the mine and exploit deeper ore bodies while employing white miners. Both of these points allowed it to distance itself from the stigma of being an inefficient, Mexican-owned mine worked by Spanish-speaking peoples. In this orgy of production, workers from all groups were injured and killed, with the Spanish-speaking workers suffering disproportionately.[14]

When the New Almaden Mine was taken over by the Quicksilver Mining Company (QMC) in November 1863, the new management introduced significant changes in the landscapes of work, and the new American owners took the racialized workforce they inherited and reinforced it, instituting new rules and regulations that although subjugating all the workers, impacted the Spanish-speaking workers most severely.[15] The new owners made radical changes to both work landscapes and the landscapes of camp life. These changes were representative of the very different ideologies brought to New Almaden by the New American owners. These changes systematically held back the Spanish-speaking workers while enforcing an even more hierarchical workforce.

In work life the new mine operators attempted to increase regimentation and to cut the cost of labor. The QMC wanted to pay workers once a month, rather than twice a month, and it insidiously attempted to make it more difficult for workers to measure and keep track of their work themselves so that they would be at the mercy of the mine company when it came to determining pay.[16] For the long-time Spanish-speaking miners and their families these changes were radical. For the English-speaking miners, who were either hired in the later years of Barron, Forbes & Co. ownership or hired in large numbers soon after the change in ownership, these changes were not so much changes as simply the way things were. The English-speaking miners were, as well, in a privileged position under QMC management, making significantly more money than their Spanish-speaking counterparts.[17]

The Spanish-speaking workers responded to the QMC changes with five strikes, three between November 1864 and January 1865, and then one each in March and April 1866.[18] The English-speaking miners did not identify with the Spanish-speaking workers and mostly stood aside in their conflict with management. With these strikes, the Spanish-speaking workers regained some ground, but the ground that was lost was of greater consequence. The QMC had succeeded in subjugating the Spanish-speaking workforce, and every day the company further reinforced the Spanish-speakers' subjugated position by hiring more English-speaking workers. As 1865, 1866, and 1867 were the years of greatest production in the mine's history, more production meant more jobs, and as more and more non-Hispanic whites were hired to fill these jobs, the power of Spanish-speaking workers decreased further. The racial, ethnic, and class differentiation of the mining landscape employed by the QMC worked both to define the manager-engineer and the laboring classes and to divide the laboring class internally along racial and ethnic lines. With a workforce divided against itself, and with workers who worked only with those who shared their eth-

FIGURE 4.9 Spanishtown, New Almaden Hill, 1880s
This photograph shows Spanishtown from Mine Hill. The Catholic church and the plaza are near the center left of the image. The main planilla and Englishtown are off to the bottom left. (Courtesy of History San Jose Research Library, Historic Images of the New Almaden Quicksilver Mines, 1979-251-136.)

nicity, the company successfully defended itself from the possibility of further mass strikes, and indeed there was never another strike by workers at New Almaden.

In the late 1870s there were five active mine shafts located on Mine Hill, all feeding cinnabar to the reduction plant at the Hacienda, built in a long, narrow valley below Mine Hill.[19] For the most part the reduction workers, mine managers, and office staff lived in the Hacienda village. The miners and *planilla* workers lived on Mine Hill, over 1,000 feet higher and three miles by company road from the reduction works, in either Englishtown or Spanishtown, from where they walked to their work at the shafts or tunnel entrances (Figure 4.9). The New Almaden Mine had about 500 employees (supporting a community of about 1,600 people total) who worked the mines and reduction plant in two 10-hour shifts, from 7 a.m. to 5 p.m. and from 7 p.m. to 5 a.m. The mine employed about seventy-five managers and staff.

Of the roughly 400 miners and laborers, 60 percent worked underground, 25 percent worked aboveground, and 15 percent worked at the reduction plant.[20] The Hacienda reduction plant and the mines on the hill were separate worlds with little interaction. Mine activities were coordinated from the hill office, and reduction activities were run from the Hacienda office. The Hacienda office was also the main office where mine officers and their staff kept track of the mine and reduction works as a whole, but few other mine employees, be they management or laborers, had any reason to cross from one world to the other.

The organizational structure of contract mining was a fundamental principle of life at New Almaden—everyone had to understand and play the contract game.[21] In contract mining, groups of miners calling themselves companies bid on work as the mine managers specified it. Every two weeks the mine advertised the contracts it was opening to bid, and the mining companies (with from four to twelve members) bid against one another for the contract.[22] These "companies," appearing most often under the name of a leader (such as "Martínez and Company") in the mine records, were organized by the miners mostly based on race and ethnicity and sometimes by family group.[23] Individual miners who came to New Almaden had to affiliate themselves with companies.[24] The miners in a company could shift frequently as men joined or quit (or were solicited or forced out) as situations warranted. Although some mining companies were partnerships of relatively equal miners, others were commanded by a boss or two who profited from the labor of those they hired.

Knowledge of the mine was critical to everyone involved; this knowledge was shared, whispered, or held secret as situations demanded. Mine managers had to know the mines to write the contracts; they learned both from supervising the work of the mining companies and, most important, from their closely and diligently kept mine records and models in which they tried to trace and predict where rich ore would be discovered. The miners had no such record, only their experience doing the work and what they could learn from other miners. Mine managers also had to know the miners and their companies. Although in theory any company of miners (even from outside the community) could show up and win a contact, in practice there were never more than a few companies actually bidding on any one contract. Savvy managers anticipated which companies would be bidding and tailored contracts to their own advantage, for example by contracting suspected difficult work to unaware companies. One mine manager at New Almaden argued that the mine company preferred contract mining because "under the contract system there is natural selection and weeding out going on among the employees, the less competent miners and dissolute miners not being able to compete with the more skilled and better class of miner."[25] Mine managers also claimed that in a mine as large as New Almaden, with miles of underground workings and teams of miners working at far-flung points, it was impossible to do more than an occasional checkup on miners' progress. In contract mining, management argued, the miners were responsible to themselves.

Mine managers stated that for mine workers, the contract system offered a ready stake in the success of the venture. An unexpected strike of rich ore could mean a huge payday.[26] In practice, however, the contract system meant that workers could not rely on a regular paycheck, but were instead at lean times forced to take advances with interest from the company and the company store when the contracts were hard and pay was low. The contract method also meant that workers had no means of comparing their rate of pay with the pay of other workers, and thus could not organize for wage increases. Under the contract system pay differences could be attributed to a combination of luck and skill, and not to the actual value of an employee's work for the mine.

Although the mine employed nearly 150 day laborers who were not officially part of the contract system, in actuality the jobs that they did—from blacksmiths who made and sharpened tools, to laborers who sorted and graded ore, to chargers who loaded the furnaces—were directly dependent upon the contract companies whose work their own jobs supported. Many workers also moved between contract work and day labor, necessitating the maintenance of allies in mining companies, no matter their current employment.[27] Outside of the employed population were family members who picked up information as they went about their day and who could facilitate company arrangements on behalf of miners in their family.

Over the years the New Almaden managers expanded the contract system until, by the late 1870s, the effort of mining the ore and then transporting it from Mine Hill to the reduction works was organized into a series of seven sequential contracts, with five governing the underground work—yardage, drilling, "tributing," "tramming," or skip loading—and two governing transporting the ore aboveground—teaming and rail operating.[28] These seven contracts organized the means by which ore was mined and then moved to the furnace yard as one contract company handed off its work to the next. The underground contracts were by far the most important; it was here that most of the work was done, and the underground work was the largest expense for the mine company. The aboveground contracts were important but modest by comparison.

The contract system divided the workforce by race and ethnicity, decreasing the opportunities for workers to communicate and share knowledge. Mine managers at New Almaden had strong views that race and ethnicity determined which kind of work a person could do best. These views were based on the skill sets that miners demonstrated historically and that then were naturalized over time. Racist perceptions

FIGURE 4.10 Company of Cornish Miners Working a Yardage Contract, early 1880s

This photograph depicts a company of yardage contractors working at the face of a tunnel. Yardage contractors drilled and blasted the tunnel face and set in place the heavy timbers. Captain Harry, the captain of the mine, is the bearded man at right. These men would have worked by candlelight: a candle is visible in front of Captain Harry. (Photograph is by Robert Bulmore. Frank J. Sullivan and Charles N. Felton, *A Contested Election in California*, between pages 11 and 12.)

about mental and physical attributes of various groups of workers also influenced the understandings of the mine managers.

Underground Work

The underground landscape of the mine, like any landscape of work, was the scene of struggles for power between management and worker groups. The underground landscape was also a mysterious place; all was darkness, there were little-known recesses everywhere, and occasionally people disappeared. Mine management attempted to survey and map the mine; workers knew it by experience. At New Almaden the management exercised its power over the underground by controlling what went in and out of the mine, while leaving many details of the underground world to the workers and their own codes of conduct.

The first act of mining was getting to the ore, accomplished at New Almaden by the first of the seven sequential contracts, the yardage contract. A company of men with a yardage contract cut drifts, shafts, winzes, and raises in an effort to extend the mine to reach new ore deposits. Yardage companies were paid, as the name implies, by the yard of rock that they removed in extending the mine. This work did not involve extracting any ore unless it was found unexpectedly. The contract did, however, involve timbering the tunnel by adding massive wooden supports where the ground was unstable. In a photograph by mine accountant (later superintendent) Robert Bulmore, a group of Cornish miners pose at the working face of a drift (Figure 4.10).[29] As a company these men drilled into rock, making holes in the mine face for explosives. After the blast the men removed the rock, placed timbers as necessary based on the geologic conditions, and began drilling again. Yardage contracts were conceived by mine management based on precise surveys and expert opinions on where ore was. The mine set standards for the quality of yardage work, although for the contract miners, temptations were ever present to cut corners. Yardage contract work was often called "dead-work," because for the mine owners it involved spending money while not extracting any ore. It was often exploratory in nature and frequently failed in the goal of opening and developing new ore bodies. Yardage contracts were most often taken by Cornish miners. Superintendent Jennings, in 1887, testified that the Cornish miners "have been the most successful men in making wages under the yardage system . . . in open competition." He continued, echoing most probably the common thinking among mine managers: "It may be accounted for by the fact that their ancestors were miners, and they were brought up in Cornwall on a system of contracts somewhat similar to what is carried on here."[30]

The next two contracts, drilling by the foot and tributing, were for actually extracting ore from an ore body, and both contracts took place in the laborés (the ore bodies shown as shaded zones in Figure 4.5). "Drillers by the foot" were paid by the length of the drill hole they hammered into an ore body for the purpose of setting an explosive charge. This work was done under the supervision of a shift boss, who measured and reported the length of the drill holes. More common in the laborés, however, were tribute contractors who were paid by the amount of each grade of ore they extracted (Figure 4.11). The tribute miners blasted and excavated the ore, and then handed it off to the trammers. The tribute mine contracts tended to be worked by Mexican miners; mine managers claimed that Mexicans had an uncanny ability to find veins of ore and to follow them wherever they led. In Superintendent Jennings's opinion, again naturalizing tendencies among worker groups, Mexican miners had been doing this kind of work at New Almaden since the earliest days of the mine.[31]

FIGURE 4.11 Tribute Miners Working in a Laboré (large ore body), early 1880s

Although these miners appear Cornish and were posed by the photographer, documents show that most of the tribute miners at New Almaden were Mexican. The men at the rock face work to extract seams of cinnabar while other men in the company crush rock and separate the ore before handing it off to the trammers. Captain Harry is seated left of center. At upper left a man with a saw works on timbering. (Photograph is by Robert Bulmore. Sullivan and Felton, *A Contested Election in California*.)

The trammers, the fourth group of contractors, lived in the world of the miles and miles of dark mine tunnels (Figure 4.12). Their job was to take the ore from the tribute company, load it into ore cars, and then push and pull these ore cars through candlelit tunnels and abandoned laborés on often long, winding paths to the vertical shafts, where cars were lifted or sometimes lowered by hoists to the appropriate level with a mine exit.[32] In Robert Bulmore's photo, the trammers are shown with a full car of ore in a laboré with the dark tunnel beyond, and a spur line to the right. Candles, at the center of the sides and top of the image, cast a faint glow to help the trammers avoid obstacles such as the massive redwood support timbers. After pushing the carts through the dark tunnels, the trammers deposited ore next to shafts in "plats," or assigned areas for each of the tributor companies.

FIGURE 4.12 Trammers in the New Almaden Mine, 1880s
Trammers pushed and pulled ore carts through the dark tunnels of the mine to the shafts. Although many trammers were Swedes, Mexican men also worked tramming ore. The location of this photo can be found (at least approximately) on Figure 4.4. Look for the 1,500-foot level of the Randol shaft where tunnels branch. (Photograph is by Robert Bulmore. Sullivan and Felton, *A Contested Election in California*.)

According to Jennings, tramming and skip filling (the fifth underground contract) were done by the Swedish, who in his estimation did not have great mining skill but who did have great physical endurance.[33] The skip fillers took the ore from the plats and loaded it into skips, large buckets that were used to haul ore up (and occasionally down) the shafts. The work of the trammers was estimated by the ton from the skip loads hauled to the outlet tunnel. The skips were raised and lowered by steam engines operated by engineers who worked by the month for the company.[34]

These first five contracts—yardage, drilling, tributing, tramming, and skip filling—were the underground contracts at New Almaden, and as such represent all the miners and the majority of workers on the hill. What becomes apparent from descriptions of work at the mine, and mine records, is that the race and ethnicity of workers were a determining factor in the kind of work that a man did. Cornish miners had yardage contracts building the mine infrastructure. Mexican miners worked in the laborés extracting cinnabar. Swedish trammers moved the ore from the laborés to the shafts, and Swedes then shoveled the ore into skips, which were raised through the shafts on their way out of the mine.

The sequential contracts meant that each contracting company was dependent on the other companies for their pay. For example, trammers could move only whatever the tributors had extracted. As in any "line" work, a person's job was dependent on the person before him doing his work, and the person behind him doing his. However, there would have been little direct communication between these groups. Cornish miners and Mexican miners often could not speak the same language. Swedes probably learned English, but as trammers and skip fillers they most often dealt with Spanish-speaking workers. Knowledge and communication were the keys to power in the mines, and the underground contract system tilted this power toward the mine management.

All contracts were not created equal, just as all racial and ethnic groups were not treated equally. Crucial to all workers was the means by which their pay was determined. In the underground contracts, the yardage companies—most often Cornish miners—and the trammers and skip fillers—Swedes—could measure their own quantity of work. Cornish miners could simply measure the tunnel they had dug, and Swedes were paid by the number of tons they loaded into cars and skips. The tributors, most often Mexicans, could not measure their own quantities of work, but instead relied upon quality and quantity determinations made aboveground. Tributors had their ore sorted, graded, and measured on the planillas (sorting floors) at the mine outlets.[35]

Aboveground Work

The most important aboveground work at New Almaden was at the planilla, where workers earned a daily wage and were under the direct supervision of mine management.[36] When the skips, traveling up or down a shaft, reached the outlet tunnel for that shaft, they were dumped directly into ore cars that were run out to the planilla. A large level area near the major ore exits from the mine where the ore was cleaned and sorted, the planilla was the transitional space between the underground and aboveground, and it was where major power struggles at the mine were negotiated. The company used three tools to control what went on in the mines: determining the contracts, awarding the contracts, and measuring the product at the planilla. This organizational structure meant that the product of the work performed under each of the awarded contracts had to be kept separate and accountable to the mining

FIGURE 4.13 Original or Main Planilla beneath Mine Hill, 1863
This is the original tunnel into Mine Hill and the original planilla. Notice the separate piles of ore at the bottom right and the discarded waste rock dumped down the hill, extreme bottom right. Compare with Figure 1.8, probably taken the same day. A part of Spanishtown is visible to the left in this photo; Englishtown was a few hundred yards to the right. (Photograph is by Carleton Watkins. Courtesy of the California History Room, 2010-0635.)

company; the planilla was where management regulated the work of all the contract workers at New Almaden (Figure 4.13). Day laborers employed by the mine cleaned and sorted the ore under the watchful eye of mine managers. The planilla was where the mine company actually took ownership of the ore, determining its quality and counting its quantity. It was also where the product of the mine, and the result of the labor of the men tributing or tramming in the mine, came out into the light.

At the Randol planilla, there were separate sorting areas for each mining company extracting ore. The planilla shed covered a long row of identical sorting and grading areas. Two Robert Bulmore photos, an exterior view and an interior view, show the Randol planilla a few hundred yards downhill and to the southeast of the Randol

FIGURE 4.14 Exterior of the Randol Planilla, 1880s
Under the long planilla roof, there were ten to fifteen identical areas for sorting and grading the ore of different mining companies. The Buena Vista shaft is shown in the center left. The valley beyond is today southeast San Jose. The shed is oriented roughly east-west. This view is looking east. (Photograph is by Robert Bulmore. Sullivan and Felton, *A Contested Election in California*.)

shaft (Figures 4.14 and 4.15). In these carefully staged photographs men perform each of the sorting and grading tasks.

In the exterior view, with views out over the Santa Clara Valley, men push ore cars on elevated rails from the mine under the planilla roof. In the center distance, a man dumps a car of waste rock. On the right side of the image, men with wheelbarrows take the sorted and graded ore from the many sorting areas under the shed and combine the ore into piles in the yard, from where it was then loaded onto wagons for the trip to the tramway down the hill to the Hacienda furnaces.

In the interior view, a man pushes the ore carts through the planilla shed on tracks fifteen feet or so above the floor, in the peaked portion of the roof. The ore is dumped over a grizzly, an inclined plane of parallel metal bars that allows ore smaller than a

FIGURE 4.15 Interior of the Randol Planilla, early 1880s
Ore from the tributing companies is being sorted and weighed. Like other Bulmore photographs, this picture is highly staged, with men shown simultaneously performing all of the tasks that went on at the planilla. This view is from the east end of the shed looking west. (Photograph is by Robert Bulmore. Sullivan and Felton, *A Contested Election in California*.)

certain size to fall through, while large chunks of ore slide down into a separate pile. Men on the planilla floor break ore into appropriate sizes, separate waste rock from good ore, and put the ore into piles corresponding to the grade of ore. Men then shovel the various ores into wheelbarrows, have it tallied, and take the ore out to the yard.[37] This sequence of tasks allowed the mine company to sort out waste rock, grade the ore for quality, and then weigh it for quantity. It was these measures that determined the rate of pay for the tributors.

On the planilla floor, there was a power struggle over the all-important measurements; floor supervisors, such as the man in the lower right in Figure 4.15, kept the tallies.[38] As part of the power struggles, the mining companies had representatives working on the floor acting in their interests.[39] The planillas were where useful work was done in the process of transforming ore to mercury; more important, however,

FIGURE 4.16 Exterior of the Randol Shaft, 1880s

In this photograph teamsters haul timbers, probably redwood, to staging areas outside the shaft. The buildings shown here housed the engines running the shaft machinery. Ore that was lifted in the shaft was run out to the Randol planilla, located downhill from the extreme bottom right of the photo. The Buena Vista shaft is visible in the distance at bottom right, a few hundred feet lower in elevation. (Photograph is by Robert Bulmore. Sullivan and Felton, *A Contested Election in California*.)

they were the primary spaces at which workers (especially Spanish-speaking) and management groups negotiated their relationships.

Moving the ore from the planilla to the furnaces involved two groups of contractors: white teamsters and Mexican or Irish rail operators.[40] Although the men working these contracts performed crucial functions at the mine, they represent a very small number of the total employees of the mine. The teamsters moved the ore in wagons from the planilla to the ore bins at the incline tramway, and they moved supplies, such as timber, on the hill (Figure 4.16). The work of the rail operators included loading cars from the bins and then running the railroad, which was designed so that the descending full cars were used to raise the empty cars from below. At the bottom of the incline, ore was dumped as appropriate into one of ten bins holding different types of ore designated for the different furnaces. There was also an additional sorting function at the bins; the *tierra ore* (fine ore that was often low-grade) was again passed over a grizzly, this time with slats closer together. The ore that could not pass through was called *granzita*. This ore tended to be slighter richer than the tierra ore as a whole. By performing this grading at the furnace yard as opposed to the planilla, the mine company avoided having to pay the tributing companies (again mostly Spanish-speaking workers) a higher price for this higher grade of ore.[41]

The contract system as an organization of work at New Almaden meant that mine workers were divided into small groups by race and type of contract. Negotiations between the mine managers and the workers took place at the scale of the contract, often giving the mine managers an advantage. Mine managers could selectively give groups advantages over others. For instance the Brass Wire Company, a group of Cornish miners working yardage contracts, colluded with a mine manager to make more money for themselves by strong-arming other workers off of well-paying contracts and by cheating the company. The power of this group was described in a letter that mining captain White wrote to the head of the mine, J. B. Randol:

> It is a well known fact that there exists a ring, the Brass Wire Company, consisting of husbands, sons, etc. . . . There are favors that can be done in many ways . . . It is a remarkable thing that Brass Wire members are never stopped on good contracts . . . others are . . . and it is not surprising that the ring should conspire to blacklist anyone likely to assume control or interfere with their freedom and management of work amongst themselves.[42]

In spite of such abuses, the contract system proved an effective tool for mine management because it gave managers a racialized workforce that could be treated unequally. Because of the hard racial and ethnic lines among workers, they were not able to organize as a class of workers against mine management.

Work at the furnace yard was accomplished by men who were paid by the day or by the month, as opposed to by contract (Figure 4.17).[43] Monthly workers tended to be highly skilled whites who supervised the operation of the complex reduction technology, whereas the daily workers tended to be the lowest-level men on the racial hierarchy, and it was these men who were most subjected to mercurial poisoning. Contract work was not used in the furnace yard, because labor could be effectively supervised (as opposed to men working in remote spots in the mines) and the furnace systems were complex and idiosyncratic, requiring men familiar with their operation.

In the furnace yard the ore from the mine was fed into one of as many as eight separate furnace systems; a furnace system included the furnace and the condensers. Reduction was the process of extracting mercury from cinnabar, and at its most basic

FIGURE 4.17 The Reduction Works at New Almaden, 1875

This photograph was taken at a time of major transition at the furnace yard. New continuous furnaces are under construction or have recently been built at the right rear of the furnace yard. Most of the older, intermittent furnaces were replaced by the early 1880s. Within the yard are stacks of firewood for the furnaces. Compare this photo with the plan in Figure 4.18, drawn only a few years after this photograph was taken. The plan shows the furnace yard reconfigured for the reduction of low-grade ore. Mine Hill is to the right; the Hacienda village is off to the bottom left, behind the photographer. (Courtesy of History San Jose Research Library, Historic Images of the New Almaden Quicksilver Mines, 1979-251-106.)

the process required two steps. The first step was to place the ore in a furnace and heat it to around 650°F, at which temperature the mercury in the ore vaporizes, creating mercury gas. The second step was to cool the gas until it condensed into liquid mercury. This was achieved by forcing the gaseous mercury through a system of airtight chambers, or condensers, which allowed heat to be released while containing the gas. Once condensed, the mercury was collected and bottled. In practice this basic process had many component steps that were modified and changed over the history of the industry.

Around 1880 the furnace yard at the New Almaden hacienda was an intricate tangle of eight separate furnace systems (Figure 4.18). These furnace systems were the production engines of the mine, and mine managers kept precise statistics on the operation and productivity of each. The furnace yard was first developed by Barron, Forbes & Co. on a level area next to a creek in the narrow valley below Mine Hill. As the village of New Almaden expanded into the rest of the valley, the furnace yard was hemmed into the end of the valley. Over the years constant modification and rebuilding of the furnace systems—often in the same place as older systems—resulted in the tight tangle of furnaces, condensers, rail lines, storage bins, and ore dumps. The bins as well as the furnace structures were all roofed over to protect them from rain, whereas the sides remained open for ventilation and cooling. The furnace yard was served by an infrastructure of roads, rail lines, and water systems, by which the furnace yard ingested and eliminated as it produced quicksilver twenty-four hours a day. The eight furnace systems operating in the early 1880s were a mix of old and new, local and European technologies, a compromise of historical circumstance and efforts to adapt technology to the changing ores coming from the mine.[44] Most significantly, the different furnace technologies were developed and employed in direct relation to the system of contract mining and the types of ores that system produced.

Most of the workers at the furnace yard, who numbered between forty and fifty, were assigned to tending one or more of the furnace systems. During normal operation a continuous-feed furnace system (the most common system in the industry after 1875) required a charger, who charged the furnace with ore and oversaw its operation; trammers, who brought ore to the furnace; and slagmen, who both discharged the furnace and tended its fires (Figure 4.19).[45] The furnace-yard trammers (most often Mexicans) had to manage ore cars on the intricate rail lines tangling over, around, and through the furnace yard. They loaded their cars at the designated ore bin, or if working a tierra or granzita furnace they might have loaded their cars at the large storage sheds for these ores. Granza (large chunks of rich ore) was precious and was processed quickly, whereas tierra and granzita ores were stockpiled in roofed sheds for processing in the wet winter months when transporting ore off the hill was often difficult. Trammers used animal power, but most often their own power, to negotiate the tracks to their assigned furnace. The continuous-feed furnaces—sometimes over forty feet tall—were fed from the top, requiring the trammers to use one of three water-powered elevators to lift the ore cars to the overhead tracks. Trammers dumped the ore from their cars, and a small amount of coal, into hoppers at the top of the furnace; the chargers then managed the flow of ore into the furnace.

Chargers (most often non-Hispanic whites) minded the performance of the furnaces, a task requiring great experience and a feel for optimum performance. The

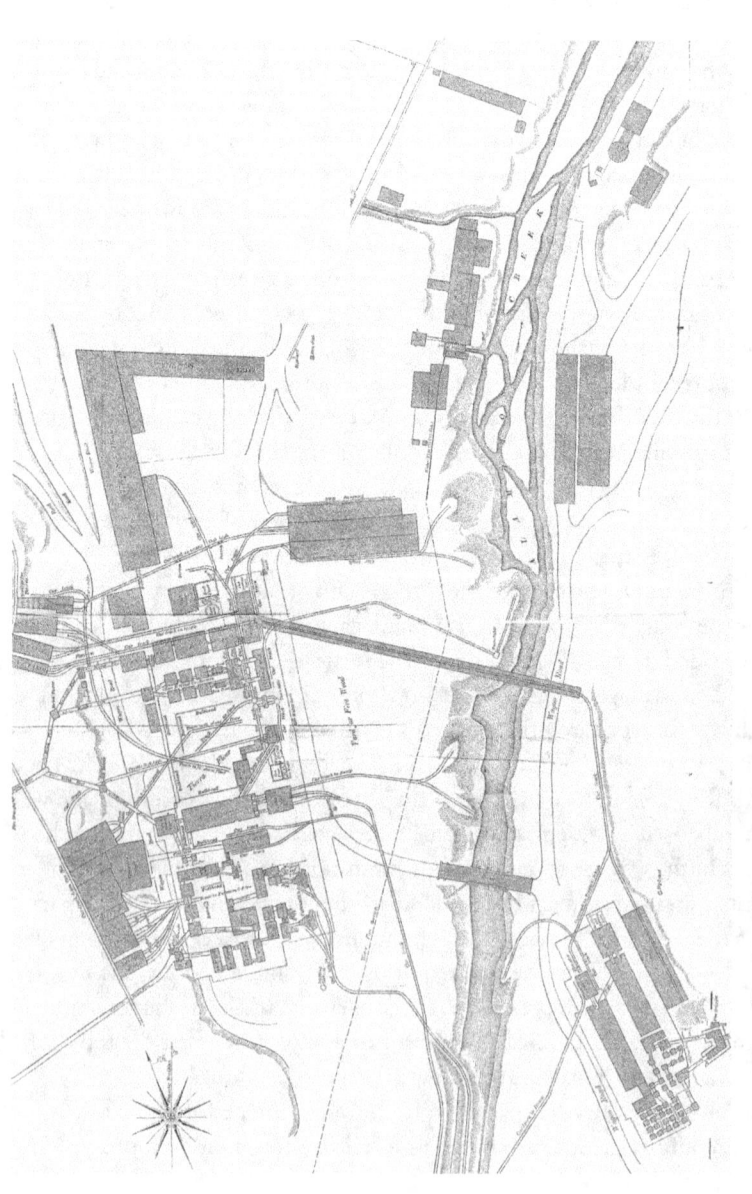

FIGURE 4.18 New Almaden Furnace Yard, 1879

The furnace yard was a complex tangle of eight separate furnace systems, each composed of ore paths leading from the ore bins to the furnaces, from which spent ore was dumped and the mercury was bottled ready for sale. (Samuel Christy, *Quicksilver Reduction at New Almaden*.)

FIGURE 4.19 Drawing Spent Ore from a Scott Furnace

Drawing spent ore was one of the most dangerous jobs at a quicksilver mine. When removed, the ore was still hot and releasing mercury gas. This man filled the ore car and then pushed it to the tailings dump. (Bradley, *Quicksilver Resources of California Bulletin No. 78*, 45.)

only input of power to these furnaces was heat from wood, and often a bit of coal or coke.[46] Chargers had to tailor the heat input in response to air temperature and humidity, the particular characteristics of the ore in the furnace at the moment, the type of wood being burned, and a host of other factors, such as production quotas to be met. The furnaces expanded as they heated and contracted as they cooled. The chargers had to watch the elaborate wooden and metal grid of furnace "stays" that surrounded the furnace and held the bricks together and that needed to be adjusted for changing temperatures. When cool, one of these massive furnaces required a week or more to reach operating temperature. The chargers also watched for an even flow of ore through the furnace. At times, ore would become jammed between the shelves. When this happened, the chargers jammed metal rods through pigeon-holes on the exterior of the furnace to dislodge the ore on the interior. Scaffolding was erected around the furnace to allow men access to all furnace surfaces for this purpose.

FIGURE 4.20 Furnace Number 2 and Condensers (front) and Furnace Number 1 (behind), New Almaden, 1863

The devastating impact of mercury on plant life is evident in this photograph by Carleton Watkins. The furnaces are under the roofs on the left side. The large brick section is a series of condensing chambers that served to cool the mercury gas coming from the furnace. The gas eventually worked its way up the pipes on the hillside toward the stack. In this long-duration exposure, gas leakage is evident from the junction box on the hillside just behind the condensers. (Courtesy of the California History Room, 2010-0639.)

Tending to the fuel and elimination needs of the furnace were slagmen, most often Mexicans. On the Scott furnaces, for instance, a slagman added wood every hour to the fireplace at the bottom. Then, every forty minutes he drew spent ore out of each of the eight draw pits on the sides of the furnace. Drawing ore was a nasty job, one that often injured the workers. Although the ore had passed its period of greatest heating, it was still quite hot and releasing mercuric gas by the time it reached the bottom of the furnace. Workers attempted not to breathe fumes coming from the hot ore, at times tying wet bandanas to their faces in an attempt to filter the air they

FIGURE 4.21 Bottling Quicksilver, New Almaden, 1880s
Mercury was sold by the flask. The flask itself was iron and weighed about twenty pounds. Into this was added 76.5 pounds of mercury, a quantity brought down from Roman measure. This man scooped mercury from a trough inside the lockable box beside him. Mercury flowed by gravity in a closed channel from the condensers to this locked box. Being both dense and valuable, mercury could be stolen and easily smuggled; thus it was closely guarded at all times. (Photograph is by Robert Bulmore. Sullivan and Felton, *A Contested Election in California*.)

breathed. Once the ore was in the cars, the slagmen pushed the cars over to the creek and dumped the ore.

These reduction furnaces, regardless of type, were like living and breathing beings, exhaling mercury gas and sulphur mixed with the products of combustion, in this case wood ash. The fires of the furnaces pushed this deadly mix with great force out of the furnace and into the condensing system. The condensing system was built to both slow the force of the draft coming from the furnace and cool the temperature of the deadly mix. The draft was slowed by having a series of chambers with small openings between them, and the gases were cooled by the large radiating surfaces. Most often the condensers were built of masonry lined and sealed with concrete. However, at

FIGURE 4.22 "Life-Saving Respirator," 1870s

Protecting workers from inhaling mercury was crucial to preventing salivation. However, there is no account in the records of furnace-yard workers actually wearing masks such as the one advertised here, if for no other reason than that the masks cost more than a man's daily pay. Further, it is doubtful that such a mask would do much good. There are multiple accounts of furnace-yard workers carrying wet sponges in their mouths and wearing bandanas in an attempt to reduce exposure to mercury gas. (Mining and Scientific Press, common advertisement in 1874, 1875.)

New Almaden many experiments were conducted on the condensing systems. Glass and wood condensers, water condensers, iron condensers, and pipe condensers were all in use at various times. At the end of their runs, the condensing systems from various furnaces were combined and routed a few hundred yards to smokestacks at a slightly higher elevation on the hillside, removing much of the remaining gas from the furnace yard, and creating a dead zone on the hillside (Figure 4.20).

As the quicksilver condensed, it ran down the sides and across the sloping bottoms of the condensers into a system of iron pipes through which it flowed to a collection vat. Every two furnace systems shared a bottling and weighing room, although the quicksilver from each furnace was collected separately (Figure 4.21). These bottling rooms were closely guarded by watchmen; bottling mercury was a prestigious job held by American or European men. Three to five pounds of quicksilver was a very small volume to conceal (about the size of a tobacco tin), yet at 40 cents a pound equaled a man's daily pay. There are accounts of men sneaking glass jars of up to twenty pounds of quicksilver from the Hacienda and selling them in San Jose.[47]

Mercury poisoning, or salivation, was always a major problem at mercury reduction plants, and the Hacienda plant at New Almaden was no exception.[48] Humans absorb mercury through the skin, through respiration, and through ingestion. A few timid protection devices, such as the respirator shown in Figure 4.22, did little to protect the workers. Gaseous mercury is most dangerous but liquid mercury—although more stable—is also constantly sublimating, giving off fumes of gaseous mercury. Mercury reduction plants improved from 50 percent efficiency to over 98 percent efficiency from the first furnaces at New Almaden in 1850 to the rotary furnaces and stainless-steel condensers of the early twentieth century. However, mercury poisoning remained a problem.[49] All mercury mines, including New Almaden, had to engage the problem of salivation.[50] Most mines placed their most expendable workers—nearly always men from the lowest racial and ethnic group employed at the mine—in the most dangerous jobs, those of cleaning the condensers and occasionally cleaning the furnaces. Other jobs in the furnace yard required only minimal exposure to mercury, and these workers, while always accumulating mercury in their bodies, could work for years without succumbing to debilitating salivation.

Race and Work at New Almaden

The landscapes of work at New Almaden were shaped by how race was negotiated at the mine. In the contract system, work was organized into discrete chunks according to a hierarchy of races. The level of control that a laboring group had over their working situation, that is, of being able to keep track of their own pay and to limit the pollution of their bodies, was largely based on their race. Rates of pay for different racial groups reflected the unequal power relations at the mine. Stephen Pitti, in his study on the history of Mexican labor in the Santa Clara Valley, created a comparison for the year 1881 of the income and deductions of workers at New Almaden based on race. He found that whites averaged $68 a month gross pay, whereas Mexicans averaged $43. Deductions ran $21 for whites but $22 for Mexicans, leaving a net income of $47 for whites and $21 for Mexicans.[51]

In addition to the employees on Mine Hill and at the Hacienda, the New Almaden Mine employed a group of Chinese men as ore pickers who, from the mid-1870s to August 1884, worked on the old rock dumps and tailings piles searching for missed bits of good ore.[52] In employing the Chinese the mine company had to be very careful to avoid inciting the other laboring groups at the mine. E. R. Sampson, in his letter to the president and directors of the Quicksilver Mining Company, New York, and dated July 7, 1869, addressed the possibility of using Chinese labor at New Almaden in the context of a general comparison between the New Idria and New Almaden Mines. Concerning Chinese miners, he wrote:

> They are asking $1.25 per day now. They are good for some kinds of work about the mine, but a miner must know something more than to drill and blast; he must understand the nature of rock, ore, and etc. They tried Chinese labor at New Idria and it did not prove successful and I have modified my views since my first report. We should have to retain some Mexican miners and some Cornishmen for timbering as they do the best work done in the mines and perhaps they might not harm us. It might be a costly experiment but I have consulted Mr. Butterworth about it and we are looking into the matter and will make an early report upon it.[53]

In large part the management succeeded in avoiding labor unrest concerning the Chinese by making the work of the Chinese men separate from the linear series of contracts that engaged all of the other mine workers at New Almaden. Although the Chinese workers found some ore, which was then sorted and added to the ore bins at the Hacienda, the quantity of ore was very small compared with the ore coming from the mine.

"Racial harmony" was an expressed goal of the mine managers. In his 1887 testimony, Superintendent Jennings stated that the chief nationalities employed at the mine were English Americans, Mexicans, and Swedes. When asked how he accounted for this, he replied:

> The proportion in which these elements have remained here is greatly due to the natural competition going on in the contract system. These nationalities also harmonize well together, for although the Cornish and Mexicans do not associate very much with each other, they agree to let each other alone, and thus have no tendency to affiliate in combinations against the management.[54]

When asked why so few Irish were employed at the mine, Jennings replied: "This element does not harmonize with the other elements here. As I have noticed, even in other mining localities, the Cornish and the Irish do not assimilate well together."[55]

The fact that mine managers endeavored to preserve racial détente (if not racial harmony) at the mine is evident in the methods of organizing work and visible in the resulting physical form of the mine. The most obvious example is the extensive sorting of ore at the planilla, a task that was rendered largely unnecessary, from the standpoint of technology, after the invention of the continuous furnaces in 1875.[56] Mining engineers visiting New Almaden commented frequently on the extensive planilla operation of sorting and grading ore in an era when other mines in the state did vastly less of this work. However, doing away with the planilla as the key control mechanism for the contract system and all that it entailed socially at New Almaden would have been revolutionary, requiring drastic changes to the whole organization of work there.

The underground section drawing of the Randol shaft at the New Almaden Mine discussed earlier (Figure 4.5) shows the extensive system of tunnels and shafts that had been developed at New Almaden by 1885. If the goal for efficient working of a mine was to create infrastructure that was just adequate (in terms of size and permanence) for the extraction of the ore, then it is worth questioning whether the underground landscape at New Almaden met the criterion of mining efficiency.[57] For nearly three decades the Quicksilver Mining Company maintained a large community of miners at New Almaden. These miners were the embodiment of mining capacity, and they needed to be actively employed to avoid labor unrest at the mine. This was particularly the case at a closed company town such as New Almaden, where the company was forced to maintain a relatively stable population of workers and had primary responsibility for their economic survival. It is not hard to imagine that mine managers created yardage contracts (the contracts most responsible for the grand shafts and tunnels) for the purpose of keeping miners working. It is also the case that mining engineers who visited the mine commented on the fact that all ore was removed from the laborés at New Almaden, not just the richer ore as was common practice at other mines. Possibly all the ore was removed in order to keep the tributors and the rest of the laborers down the line actively working, and to keep the contracts manageable. This mining capacity, combined with managers' (and possibly everyone's) grand hopes for the New Almaden Mine (that it would rival the Almaden and Idrija Mines that were worked for centuries), created the magnificent, and arguably overbuilt, underground world in Mine Hill.

Creating Racialized Camps

As with the landscapes of work at New Almaden, the landscapes of quicksilver mining camp life were fundamentally affected by struggles over race formation and the racialization of camp landscapes.[58] Based on the composition of the workforce, a quicksilver camp could have definable white, Chinese, Mexican, Chilean, Cornish, Irish, or other landscapes, and the struggles between these groups had physical forms. The camps are best understood as sets of distinct racialized landscapes, tied to specific overlapping racial and ethnic groups, with the quicksilver camps as the sum of all these racialized landscapes.

Aside from the missions, the Hacienda Nuevo Almaden was the first and most significant Spanish colonial settlement in Alta California. It was a large settlement with many hundreds of people, all dependent on a single employer—the New Almaden Mine—and all living on company property.[59] By 1851 the settlements at the mine were populated by a few hundred miners and laborers, their families, the mine managers and their families, and providers of services (peddlers, barbers, water carriers, etc.) and their families.

In gold rush California the New Almaden Hacienda was unique, due to the particular demands of mining quicksilver and due to New Almaden being developed by the British imperialists Barron, Forbes & Co. As discussed in chapter 3, mining mercury involved significant development capital and an industrialized, wage labor workforce. The extensive and permanent camps at New Almaden were constructed and organized to reproduce this workforce, and stand in contrast to the camps of the gold rush, which were agglomerations of temporary structures built by individual miners on public land.

Barron, Forbes & Co. developed New Almaden based on the company's experience managing silver mines in Mexico and elsewhere, where they employed groups of Mexican and other miners. As a frontier outpost populated with workers recruited from Mexico and Chile, New Almaden was a hacienda where the workers were wholly dependent on the company and the company on the workers. In 1851 there was no other place in the state like it; moreover, as California quickly developed in the gold rush, New Almaden became even more of an anomaly, a British/Mexican island in a rapidly Americanizing California.

The camp landscape established at New Almaden in the early days of Barron, Forbes & Co. ownership formed an armature of camp life at New Almaden that persisted for the life of the mine. There were distinct camps at the mine: the village in the valley below Mine Hill, and a camp on the hill near the mine (see Figure 1.7).[60] The Hacienda was for the European and American mine officers, whereas the camp on the hill was for the Spanish-speaking workers (Mexicans, Chileans, and Californios) and came to be known as Spanishtown. The distinctions among the camps were by race and ethnicity, more than by class, with different rules and landscape forms based on the race and ethnicity of the inhabitants.

The Hacienda, built in the long, narrow valley, was the formal show-village of the mine, composed of the superintendent's house, the Casa Grande; the company-built residences for mine officers; and other company buildings, including a company store, a butcher, a grocer, and a hotel for visitors. The Hacienda was a street village built along the road from San Jose that ended at the reduction works (see Figure 1.5). Linear design elements added cohesiveness to the settlement. From the Casa Grande at one end of the Hacienda to the reduction works at the other, the road was lined with sycamore trees. The creek running through the valley was partially diverted

FIGURE 4.23 Casa Grande, New Almaden, 1880s
Over its five decades as the mine superintendent's residence, the Casa Grande had many facelifts. This particular version is from the 1880s. The Chinese pagoda is at left. (Photograph is by Robert Bulmore. Sullivan and Felton, *A Contested Election in California*.)

to create a small stream that ran between the sycamore trees and the sidewalk. The company-owned houses, mostly four-room wood-framed structures, were carefully spaced along this road, each with a front garden surrounded by picket fencing.

For the visitor to the mine, or for the superintendent making his trips between the Casa Grande and the mine office at the reduction works, the Hacienda village was a long promenade, carefully designed to create a striking visual impression. William Brewer described the Hacienda in 1861: "A most lovely town has sprung up by the furnace—neat houses on a long street, with a row of fine young shade trees, green yards, pleasant gardens, etc."[61] For Hacienda residents the pressure to maintain appearances must have been intense, as their houses and gardens functioned as the image of the mine. Years later Mary Hallock Foote felt this power in the landscape.[62] She described the Hacienda in an article published in 1878:

> The charms of the Hacienda are of the obvious kind: a long, shady street, following the bright ripples of a stream (which the tourist generally speaks of as the "Arroyo de los

FIGURE 4.24 Chinese Pagoda, Casa Grande Grounds, New Almaden, 1880s
This pagoda was presented to the New Almaden Mine during the 1850s, when the mine was under Barron, Forbes & Co. ownership. Through the nineteenth century, it was a centerpiece of the gardens of the Casa Grande. (Photograph is by Robert Bulmore. Sullivan and Felton, *A Contested Election in California*.)

Alamitos"), at one end the manager's house, with its double piazzas and easy hospitable breadth of front, a lonely background of mountains at the other, and the vine-covered cottages between. These agreeable objects can be as well appreciated in a drive along the main street as in a year's residence there,—it is very pretty; but as the "show" village of the mine, ever conscious of the manager's presence, the Hacienda wears an air of propriety and best behavior, fatal to its picturesqueness.[63]

At the entrance to the Hacienda valley was the Casa Grande, a three-story, twenty-seven-room stately brick building that was office for the mine, residence for the mine manager, and hotel for visiting guests (Figure 4.23). One of the grandest buildings in California at the time of its construction, the Casa Grande, with its extensive grounds, served as a country retreat for the dignitaries visiting the mine.[64] A Chinese pagoda, presented to the mine in the 1850s by powers in China (mine folklore says the emperor himself), who were at the time buying New Almaden quicksilver, graced

FIGURE 4.25 The Catholic Church in Spanishtown, New Almaden, two views
At left, a simple, early Catholic church is at the far end of the plaza. This is a detail of an 1863 photograph by Carleton Watkins. The photograph on the right shows the last in a series of churches on this spot, built in 1885 by the community with company aid in money and supplies. This Catholic church was the most elaborate structure in Spanishtown. The elegant, classically inspired four-bay frame structure was sheathed with clapboards and had a detached bell tower at left. (Left image: Courtesy of the California History Room, 2010-0644. Right image: Courtesy of History San Jose Research Library, Historic Images of the New Almaden Quicksilver Mines, 1979-251-136.)

FIGURE 4.26 The Center of Englishtown, New Almaden, 1880s
The residents of Mine Hill had a commanding view of the Santa Clara Valley. Company boardinghouses are in the lower right, the company store is in the center right, the mine office is at the center, and the Methodist Episcopal church is off the top right. The Englishtown school is to the left. (Photograph is by Robert Bulmore. Sullivan and Felton, *A Contested Election in California*.)

the grounds of the Hacienda for decades and was always a centerpiece in the formal garden design (Figure 4.24).

Although the road from the Casa Grande through the Hacienda was a grand promenade, it was also the main road from San Jose, heavily traveled by supply wagons and all other traffic to and from the mine. The mine company closely regulated this road like all other roads at New Almaden—company supplies and quicksilver had the right-of-way. For Spanishtown residents, the roads from the Hacienda to Spanishtown were long and steep, with a change in elevation of over 1,000 feet. Trips to and from Spanishtown were not taken casually, especially if on foot. Located next to Mine Hill was the original miners' settlement, and from the earliest days of Barron, Forbes & Co. development the village housed a few hundred workers and their families and was a settlement of significant size in early California.[65] The village

FIGURE 4.27 Cottages, Englishtown, New Almaden, 1880s
These cottages, home to the Cornish miners, dotted the hillsides of Englishtown. (Photograph is by Robert Bulmore. Sullivan and Felton, *A Contested Election in California*.)

was perched on a dramatic ridge connecting Mine Hill and Cemetery Hill, on which Spanishtown residents had their cemetery.[66] Like the Hacienda, Spanishtown was a street village, but unlike the uniformly spaced managers' cottages, Spanishtown was characterized by densely packed buildings and houses that formed a long street down the crest of the ridge from Mine Hill to its end at the Catholic church (see Figure 1.7). A steep drop in elevation from Mine Hill caused the village to have an upper and a lower section, with the Catholic church and main plaza in the lower village.

Spanishtown had a diverse collection of buildings, including mine company buildings, houses built by Spanishtown residents for themselves, houses that residents built for rental income, and company-built residences that were rented to employees. Most of the Spanishtown buildings were of wood-frame construction with pitched roofs. Miners' cabins were basic two-room board-and-batten structures with a porch on the front and an added lean-to on the back. Many Spanishtown houses were built right up to the street in front, with the front porch as an intermediate space between the street and the house interior. This is in contrast to the houses at the Hacienda that had fenced front gardens. In back the Spanishtown houses had fenced areas for performing household chores, raising chickens, and growing vegetables.

The Catholic church was the most elaborate structure in the village, an elegant, classically inspired, four-bay frame structure sheathed with clapboards (Figure 4.25). A freestanding structure at one side of the plaza, the church was a standout while most nonresidential structures (stores, saloons, halls) in Spanishtown were part of the densely packed street. Instead of front porches these structures had arcades, which like the front porches on the houses served to buffer the buildings from the street.

Later in the Barron, Forbes & Co. years, in the mid- to late 1850s, a settlement of non-Spanish-speaking employees living on the hill took form. This settlement, which came to be known as Englishtown, grew up a few hundred yards downhill and to the east of Spanishtown and the main planilla. Centered on the hill office of the mine company, Englishtown eventually included boardinghouses, a school, other company buildings, and a Methodist Episcopal church. The center of Englishtown was the company command center on the hill. The mine companies were paid at the hill office, and on paydays miners filled the central area. The main road up the hill from the Hacienda led to the central plaza, and then on up the hill to Spanishtown. The residents of Englishtown did not regularly encounter Spanishtown, but residents of Spanishtown regularly passed through Englishtown. Also, the later mine workings were developed to the east of Englishtown, on the other side from Spanishtown, making Englishtown more central and Spanishtown more marginal to hill activities.

When the Quicksilver Mining Company took over the mine in 1863, Englishtown underwent rapid development as more company buildings were built, including a company store, further strengthening Englishtown as the power center on the hill (Figure 4.26). In contrast to the dense street village form of Spanishtown, Englishtown was a collection of hillside cottages radiating out from a company-dominated center. The miners' cottages (generally larger than those of the Mexican miners) were simple wood-framed structures, each sited individually on the hillsides surrounding the center of Englishtown. These cottages all had picket fences defining gardens with ornamental shrubs, vines, and flowers and outdoor areas for chores. The native live oak trees were carefully pruned and were an integral part of the overall appearance of Englishtown. As with Spanishtown, there was a mix of worker-owned and company-owned houses, although the QMC did not allow anyone to build his own house on the hill after it took over.

The differences in appearance between Spanishtown and Englishtown are striking. In Spanishtown the most important aspect of the community was the street and the plaza; the buildings were built to define these spaces. Community life took place in

FIGURE 4.28 Payday at the Hill Office, Englishtown, New Almaden, 1880s
The employees are forming in front of the Mine Hill office to collect their pay. The company store is just visible at the far right side of the image. (Photograph is by Robert Bulmore. Sullivan and Felton, *A Contested Election in California*.)

these spaces. Most Spanishtown houses had a front porch, where one could be part of the street yet also at one's home. The Cornish cottages were carefully spaced from one another over the hillsides, with gardens and yards defined by fences. In Figure 4.27, showing cottages built on both sides of a canyon, all of the cottages are oriented the same, facing down and out of the canyon, toward a scenic view. Englishtown cottages also had front porches, but here the porches connected the interior to a fenced garden, while facing the distant view. Unlike the one main street of Spanishtown, hillside paths connected the cottages to one another and to the center of Englishtown. For residents, the center of Englishtown was a place you went to, not something your cottage was part of. Going to the center of Englishtown meant a deliberate trip.

The racialization of space at New Almaden was stark, as it was established under Barron, Forbes & Co. management in the 1850s. The Spanish-speaking workers, in their own village, were out of sight of the genteel Euro-American visitor. Spanishtown was many things to many different groups of people. To the Mexican population it was a safe haven in an increasingly unfriendly state, whereas to the larger community of the Santa Clara Valley Spanishtown was commonly seen as a haven for outlaws. In contrast, the company presence was dominant in Englishtown, with the center of town containing company offices, boardinghouses, and stores. The hillside cottages, distant from Englishtown center, may have been in part an escape from the company eye.

Racialized Paternalism

After 1863 the mine was no longer run by the British imperialists who had first developed it; instead it was run by the QMC, which consisted of Americans and businessmen from the East Coast, who had different ideas of how to operate the mine and a different set of conditions under which they were making decisions. As was detailed in chapter 2, the mine was not profitable for the new owners and they struggled for a number of years while borrowing heavily. The QMC management believed that the camps at the mine could be a source of revenue for the company. The company introduced paternalist policies at New Almaden as one method of tailoring the mine to its wishes. Paternalism is a system in which an individual or a group of people with power provides for and regulates the conduct of a group of people under their control.[67] The term *paternalism*, of course, refers directly to a father-child relationship. Whereas Barron, Forbes & Co. was comparatively hands-off regarding workers' lives outside of work, the QMC made mine company policy that directly affected workers' lives. The QMC policies, however, were not the same for every group at the mine. These policies were racialized, with distinct policies for each racial or ethnic group, due to the complex racial and ethnic structure of the population at the quicksilver mine. At New Almaden paternalism influenced the landscapes of reproduction, making paternalism a spatial relation as well as a social relation.

Historian Philip Scranton has defined paternalism as composed of three overlapping spheres: provision, protection, and control.[68] The landscape changes made by the QMC formalized relationships between management and worker groups in each of these spheres. The changes that were made affected the racial and ethnic groups at the mine differently. When the QMC took over the mine, one of the company's primary goals was to secure its new property by transforming New Almaden into a company town where the company owned all the property, controlled who lived in the town, controlled all comings and goings, and controlled all commerce. Under Barron, Forbes & Co. management, the settlements on the hill were built on company property but were mostly paid for by the mine workers, and the company was

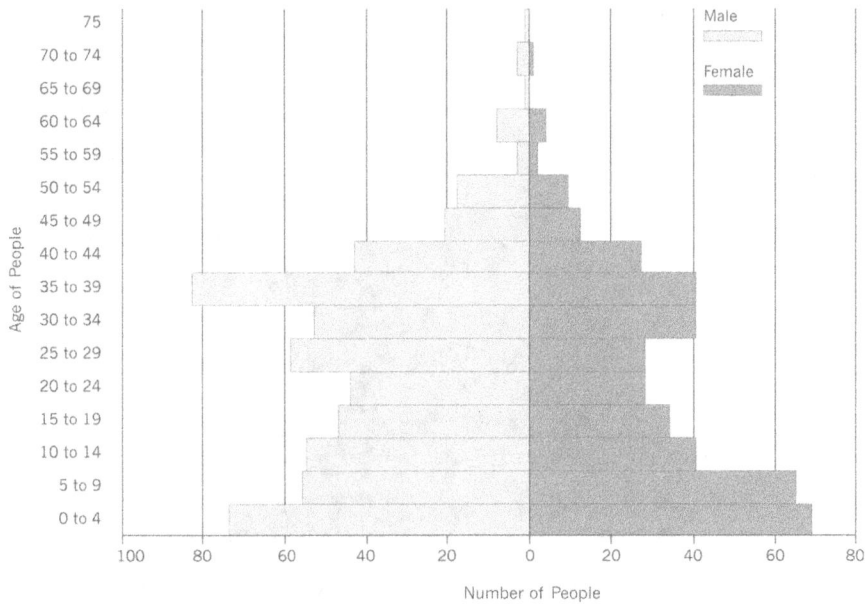

FIGURE 4.29 Population Pyramid for the New Almaden Mine, 1870
This chart combines all three settlements at New Almaden. (Based on the Records of the Bureau of the Census, Manuscript Schedules of Decennial Population Censuses, 1870, New Almaden, Santa Clara County, California.)

not interested in collecting land rent.[69] While the majority of workers on the hill did work for the company, there were many who did not. Some of these nonemployees ran stores, saloons, or restaurants or provided other services to the people living on the hill. Spanishtown, as an exclusively Spanish-speaking settlement, was also a "safe zone" in an increasingly hostile greater California and probably was a refuge for some. The changes that the QMC imposed were radical; suddenly the people who owned structures were told they had to pay ground rent, households without someone working for the company were forced to move, and independent stores and other services were forced to close.

New company stores were constructed in Englishtown and the Hacienda, both owned by Samuel Butterworth, the QMC's general agent at the mine (Figure 4.28). For Butterworth the camp stores were an additional mechanism for personal wealth, a means of compensation granted to him by the QMC board in New York. Mexican workers, however, complained that the company store—in addition to charging very high rates for low-quality goods—was not in Spanishtown, making it hard for them

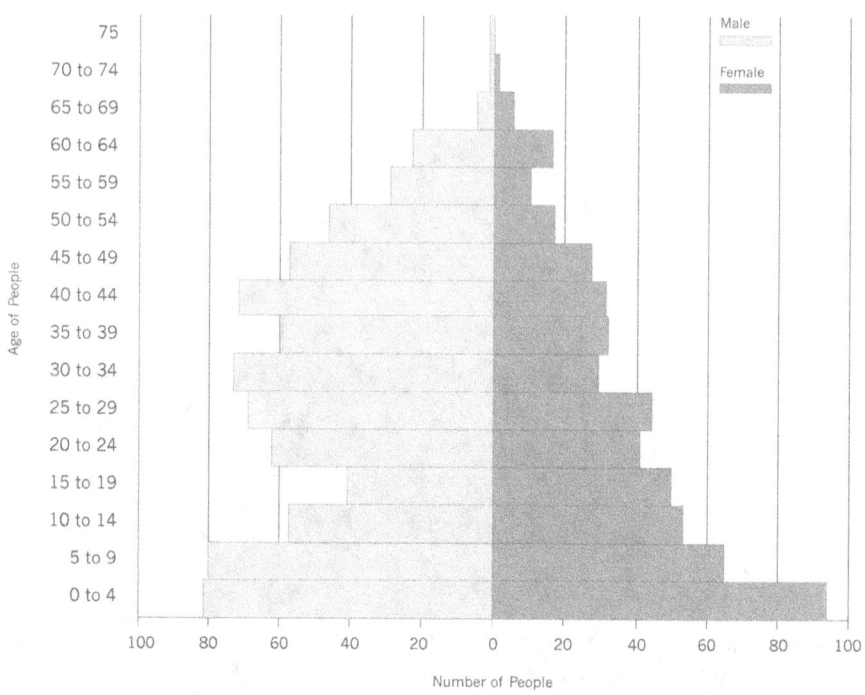

FIGURE 4.30 Population Pyramid for the New Almaden Mine, 1880

This chart combines all three settlements at New Almaden. (Based on the Records of the Bureau of the Census, Manuscript Schedules of Decennial Population Censuses, 1880.)

to get there when it was open and requiring Mexican women to do the shopping. These women, the workers claimed, were treated badly by the store owners.[70] Before the QMC takeover, peddlers had traveled door to door throughout the camps at New Almaden. The company also put up a toll gate at the hill end of the Hacienda, charging a toll to hill workers when they entered the mine and controlling the hours people could enter and leave.[71] Hacienda residents did not have to pass the toll gate to enter or leave the Hacienda. The company argued that the road to San Jose was company property. The toll gate was also used as a checkpoint where persons were searched for outside goods that by mine regulation had to be purchased at the company store.

The strikes that Spanish-speaking miners held involved these camps as much as they involved the working conditions discussed earlier.[72] These strikes were staged on the hill and included rallies in the plaza in Spanishtown.[73] But it was not all the

FIGURE 4.31 Helping Hand Club, Hacienda, New Almaden, 1880s
The Helping Hand Club is representative of Progressive Era ideals brought to the New Almaden population by mine management. (Photograph is by Robert Bulmore. Sullivan and Felton, *A Contested Election in California*.)

workers at the mine who participated in the strikes—it was primarily the Spanish-speaking workers who organized against the QMC, implying that the Cornish miners did not join with the Mexicans as a laboring class in opposition to the mine management.[74] The workers' demands concerning the camp landscapes were for reduced ground rent, an end to the company monopoly on the buying and selling of goods, and a public road to San Jose.

In their conflict the mine managers used rhetoric that conflated the Mexican protesters with immoral heathens.[75] On February 7, 1865, the mine manager wrote to one of the mine's attorneys:

> My dear Sir, I believe in Calhoun's maxim "masterly inactivity" for the present, I do not want to proceed against Garnes (Jose Garnes, the leader of the protests) until the present condition is thoroughly broken up; I want at present to drive out the vagrants and gamblers, after that is done I intend to proceed against the chief conspirators and Garnes as their leader; but until I get the leases signed I don't wish to disturb the

conspirators, ce que est différé n'est pas passée—I hope you will convict some of the gamblers and vagrants and Sabbath breakers.⁷⁶

Historian Stephen Pitti has argued that the Mexicans at the mine tied their plight to that of Mexico in its struggles against colonial powers and charged the QMC with acting as a separate government, ignoring Californian and American law.⁷⁷ The result of the strikes, in the camp landscape, was limited backpedaling on the part of the company. Land rent was enforced but reduced to two dollars every six months. The terms of the lease dictated that workers could sell their houses only to the company or to a company employee. Other results included relaxing, at least temporarily, the rules concerning the company store, and the end of hill residents being searched by the company when returning to the mine.

After April 1866 the workers never again held a strike at New Almaden. The QMC managed to divide the workforce against itself by using race and ethnicity to transcend class. Living and working with only those who shared race and ethnicity—or who were similarly placed on the racial hierarchy, as was enforced through the racialized paternalist policies—meant that class (as owner versus worker) was not a way to organize workers to better their positions. Instead the racialization of the camp landscape helped create, and then reinforced, the competition among the racially and ethnically defined groups of workers. The Cornish and Mexican populations at the mine, however, managed to maintain the family-oriented structure of the community and to resist management's attempts to incorporate Chinese men into the workforce. Census data from 1870 and 1880 for New Almaden show a population distribution that is representative of stable, family-oriented communities (Figures 4.29 and 4.30).⁷⁸ The 1880 chart shows working-age men as a larger proportion of the overall population than in 1870. The New Almaden Mine continued to support families, and although the QMC is reputed to have increased the number of Cornish miners relative to Mexican miners, in fact over this time period the number of Spanish-speaking workers also increased.⁷⁹

Work and Camp Life at New Almaden after 1870

The quicksilver boom and bust of the 1870s brought enormous changes to the mercury mining industry, particularly in the development of a handful of new major mines, a few boom districts, and dozens of prospects throughout the state. However, New Almaden—due to its extraordinary richness—retained its structure of work and camp life, with only modest modifications. James B. Randol, the nephew of Samuel Butterworth who took over the mine in 1870, serving as superintendent until 1892, instituted these modifications. Randol was a practical man who had an interest in progressive reform and who asserted his beliefs in the mines and camps. Formerly the company secretary in New York, Randol was one of a new breed of corporate manager. He was not involved in the dealings of the Bank Crowd; he instead strove to manage an efficient and profitable mine. Unlike Butterworth, who spent much of his time in San Francisco and on the peninsula, Randol lived at the mine and was heavily involved in the day-to-day operations of all its aspects.

Over his twenty-plus years at the mine, Randol instituted a range of social and benevolent societies, including schools and churches in each of the camps, Helping Hands clubs in Englishtown and the Hacienda, reading rooms, and a company doctor and medical care for one dollar a month (Figure 4.31). His scrapbooks show an interest in ideas of modern living promoted in the late nineteenth century, including the innovative idea of a living room, eliminating the formal parlor. Near the end of his tenure at the mine, in 1890, he founded technical schools at the Hacienda and Englishtown, providing training in cooking, sewing, carpentry, and blacksmithing. Many of Randol's progressive efforts at the mine were exercised unevenly. For example, of the students in the technical schools, only three or four had Spanish surnames.[80]

The Hacienda Nuevo Almaden of Barron, Forbes & Company created landscapes and social systems that remained for the life of the mine. Once established, the work and camp landscapes of New Almaden were difficult to transform, as the QMC discovered in the late 1860s. No other mine in California, or in the American West, operated on a major contract system or was racialized in this way. Large gold and silver mines, many of which were unionized, excluded racial and ethnic groups who were employed at New Almaden.

As we will see in the next two chapters, other original mines—such as New Idria and Guadalupe—were developed similarly to New Almaden, with much more complex racial and ethnic organization of work and camps than other mines in the state. The new mines, however, were able to employ mostly Chinese labor, giving those mines a unique role in mining in California and the American West.

Notes

1. For an excellent look at the distinctive nature of the underground landscapes, see Robert McCarl, *Contested Space*.
2. Tomas Almaguer in *Racial Fault Lines*.
3. Ibid., 7.
4. For information on Native Americans at the New Almaden Mine, see Coomes, "From Pooyi to the New Almaden Mercury Mine."

5. The racial and ethnic categories that I use are American (including Canadian), Cornish/English/Scottish (CES), other European, Native Californian (Spanish surname, born in California), Mexican, Chilean, and Chinese. There are no discernable blacks or Native Americans working at quicksilver mines in the census data. I have arranged these categories according to the racial hierarchy prevalent in California at the time. Concerning this hierarchy, see Almaguer, *Racial Fault Lines*. The figures presented in the charts are based on my own counts from the manuscript census and are based on the listed country of birth. Eighteen sixty may be a difficult census year for which to get an accurate count for the New Almaden Mine because the mine was closed by injunction from November 1858 to January 1861. However, through a close analysis of housing occupation in the census, I believe that most of the New Almaden miners remained on the hill during the injunction. All sides in the legal contest had a strong interest in keeping the workers at the mine both to maintain the mine and to be ready to work when the mine reopened. As soon as the injunction was lifted, the mine produced record quantities within a month and continued with record-breaking production, further strengthening the theory that most employees were on the hill during the injunction.

6. Otis E. Young Jr., *Western Mining*, 79–89. According to Young, this system was instituted during the Spanish conquest of the New World.

7. Superintendent Jennings said of the Mexican miners: "The original owners of the mine, receiving their grant from Mexico, introduced the Mexican system of mining; in this manner a great many of them came from Mexico, and their descendents have grown up at the mine, and they have made a specialty of working in ore chambers more than in competition with the Cornishmen on the yardage contracts." Frank J. Sullivan and Charles N. Felton, *A Contested Election in California*, 21.

8. Elisabeth L. Egenhoff, *De Argento Vivo*, 117.

9. For a full description of overhand stoping at New Almaden, read Frank Reader's letter to J. B. Randol, May 2, 1883, in Randol, *Papers Relating to Quicksilver Mining, Ca. 1849–1894*, vol. 13.

10. A. K. Hamilton Jenkin, *The Cornish Miner*, 204.

11. Stephen Pitti, "Quicksilver Community," 82. See also Pitti, *The Devil in Silicon Valley*.

12. Important works on this topic include ibid.; Lisbeth Haas, *Conquests and Historical Identities in California*; and Alexander Saxton, *The Indispensable Enemy*.

13. Both New Almaden and New Idria were involved in major land cases in which the Mexican land claims were not supported in the American courts. Barron, Forbes & Co. and its successors were on opposite sides in the cases involving these two mines, losing at New Almaden but winning at New Idria. For more on land and ownership in early California, see Haas, *Conquests and Historical Identities in California*; Leonard Pitt, *The Decline of the Californios*; and W. W. Robinson, *Land in California*.

14. Pitti, "Quicksilver Community," 88.

15. For an in-depth study of this topic, see ibid. and Coomes, "From Pooyi to the New Almaden Mercury Mine."

16. Their methods are explained later in this chapter.

17. Pitti, "Quicksilver Community," 82.

18. For a detailed analysis of the strikes at New Almaden, see ibid., and Coomes, "From Pooyi to the New Almaden Mercury Mine."

19. The sources for this analysis of working conditions at New Almaden about 1880 are varied. Articles by Christy provide much of the technical information on mining, particularly reduction. Samuel Benedict Christy, *Quicksilver Reduction at New Almaden*. Descriptions of the contract system and the organization of work underground include various reports in Randol, *Papers Relating to Quicksilver Mining, Ca. 1849–1894*; and Jimmie Schneider, *Quicksilver*. The best information on the experience of Mexican miners is in Pitti, "Quicksilver Community." See also Pitti, *The Devil in Silicon Valley*.

20. The numbers come from Sullivan and Felton, *A Contested Election in California*, 152.

21. Contract mining had been used at New Almaden since the earliest years of Barron, Forbes & Co. operation. See the early travelers' accounts in Egenhoff, *De Argento Vivo*.

22. Information on the contracts up for bid included "the last taker, last price, number of men required, location and size of drift and the character of the work done." Sullivan and Felton, *A Contested Election in California*, 14.

23. See Jenkin, *The Cornish Miner*, chap. 6, for more on the way that work was often organized in the Cornish mines.

24. Workers came to the mine in a number of ways: the mine advertised for workers in regional papers, and workers came to the mine through contacts with employed miners. At the mine the workers self-selected into companies, although mine management at times had influence in the composition of the companies.

25. Sullivan and Felton, *A Contested Election in California*, testimony of Superintendent Jennings, 15, 1887.

26. James Harry, a long-time miner who advanced to become mine captain at New Almaden, related the range of contract pay from his own experience. "I have worked contracts for as low as $6 and my board for two weeks. That would be $3 a week, and I have made as high as $70 and board for two weeks on contract." Ibid., 127. The Quicksilver Mining Company claimed that for 1886 and the first three months of 1887, the general average daily pay for yardage contractors (Cornish companies had the majority of yardage contracts) was $2.48. Sullivan and Felton, *A Contested Election in California*, 153.

27. See James Harry's experience in Sullivan and Felton, *A Contested Election in California*, 127, and the Anderson resignation affair, where men moved fluidly between contract work and day labor. See Randol, *Papers Relating to Quicksilver Mining, Ca. 1849–1894*, C-B 619, vol. 13.

28. This list of contracts was presented by Superintendent Hennen Jennings in Sullivan and Felton, *A Contested Election in California*, 13–15.

29. Robert Bulmore was a company accountant under James Randol who took over as manager upon Randol's retirement. As a photographer, Bulmore pioneered the use of magnesium light for his underground photography. His photographs are carefully composed images of miners, their tools, and their comrades in their underground landscapes.

30. Sullivan and Felton, *A Contested Election in California*, 21.

31. Ibid., 21.

32. At times, the most efficient exit for ore from the mine was from a tunnel entrance on the side of the hill that was below the laboré.

33. Mexicans also worked as trammers at New Almaden. Testimony of Superintendent Jennings in Sullivan and Felton, *A Contested Election in California*.

34. Mechanics and engine drivers worked by the month, whereas men about the landings and shafts worked by the day. See the comments of Hennen Jennings in ibid., 13.

35. This analysis is based on the testimony of Hennen Jennings in ibid., 14–15.

36. Contracts were less important aboveground; they organized only the transportation of ore from the hill to the furnace yard.

37. At New Almaden the coarse ore was called *granza* and the fine ore *tierras*. Before the invention of the fine-ore furnace (either the Knox-Osborn or the Hüttner-Scott) around 1875, the New Almaden Mine company paid miners only for granza because they had only coarse-ore furnaces. Before 1875 the fine ore had to be molded into *adobes*, or bricks, in order to be properly burned in the furnace. Due to the labor costs involved in making adobes, fine ore was much less valuable than the coarse ore, which could be fed directly to the furnaces.

38. See *Nuevo Mundo*, February 8, 1865, for Mexican complaints against a white manager at the planilla. From Pitti, "Quicksilver Community," 126.

39. These representatives were one of the compromises from the strikes of the mid-1860s. See ibid., 122–25. (See *Nuevo Mundo*, January 18, 1865, and January 20, 1865, for his sources.) In Cornwall a man called a sampler determined the grade and value of an ore sample. Cornish miners in Cornwall had the right to go to another sampler if it was felt that there had been an injustice. Jenkin, *The Cornish Miner*, 204.

40. This assertion is made based on job listings in the 1870 and 1880 manuscript censuses for New Almaden.

41. Testimony of Superintendent Hennen Jennings in Sullivan and Felton, *A Contested Election in California*.

42. As quoted in Schneider, *Quicksilver*, 179.

43. There were attempts to introduce a small number of contracts into the furnace yard for "grunt" work such as hauling and dumping.

44. Prior to 1867 all of the furnaces at New Almaden were of the intermittent type, fired for a few days and then left to cool for a few days. A typical intermittent furnace went through three firing cycles in a month. One of these furnaces was still in use in the 1880s (number 6 on the Jennings plan of 1879) to process "overflow" quantities of granza ore. Tierra, or fine ore, which tended to be much less rich than granza, had little value for the mine company during its first two decades of operation. But as the supply of granza ore diminished in the late 1860s and early 1870s, the mine company looked to the large supplies of tierra as the future of the mine.

45. Intermittent furnaces (common before 1875) required amounts of labor that varied over the ten-day firing cycle. At the time of Christy's article, furnace number 6 required eight men working one day to charge it. One man then had to tend the furnace for five days while it was

fired. The furnace was then allowed to cool for three days, after which it took four men one day to discharge it. Christy, *Quicksilver Reduction at New Almaden*, 559.

46. The Exceli furnaces, imported from Idrija, required coke to be mixed with the ore. For extensive information on quicksilver reduction technology, see L. H. Duschak and Curt N. Schuette, *The Metallurgy of Quicksilver, Bureau of Mines Bulletin 222*.

47. Randol, *Papers Relating to Quicksilver Mining, Ca. 1849–1894*.

48. Mrs. S. A. Downer wrote in 1854 concerning mercury poisoning at the mine: "Notwithstanding the precautions used, the escape of arsenic with the sulfate of mercury, has a deleterious effect upon those who labor among the furnaces. Each man works one week out of four, and then changes to someone else. Even cattle, if allowed to browse at large in the vicinity during the dry season, become salivated, and die from its effects." Downer, "On Her Trip into the New Almaden Mine," 117.

49. In 1886 the New Almaden Mine physician, Dr. W. S. Thorne, when asked if salivation had been a chronic problem at New Almaden, said: "No, I cannot say it has been. I have seen some very bad cases here; very bad; the worst that I ever saw; but in those cases which I treated they were cases where the men had done things that they were not required to do, and foolish things, and in fact things that they ought not to do, and we put a stop to it." His contradictory statement belies the problem. Exploitable workers could always be found to do the worst jobs at mercury reduction plants, and mine managers seldom saw any economic benefit from achieving the highest levels of reduction efficiency by eliminating escaping mercuric gas. Sullivan and Felton, *A Contested Election in California*, 39.

50. At the Redington Mine, wrote mining professor Egleston in 1875, "as soon as the workmen experience any sensitiveness about the mouth they are instructed to go to the office, and are there furnished with a mouth-wash, consisting of 2 parts of cinchona, 1 part tincture of myrrh, and 3 parts of water, but no case of real salivation except from the carelessness of the men, has occurred in a great many years." T. Egleston, "Notes on the Treatment of Mercury in North California," 284.

51. Pitti, "Quicksilver Community," 82. Deductions included advances from the company store or from the mine company and payments to the Miners' Fund. Chinese workers grossed about thirty dollars a month—however, their economy was separate from that of the other workers at the mine; they housed themselves and had their own store.

52. Sullivan and Felton, *A Contested Election in California*, 22.

53. Kenneth Rank–New Idria Mines Collection, 1867–1973, box 1559, folder 6, "E. R. Sampson to the President and Directors of the Quicksilver Mining Co., New York, July 7, 1869."

54. Sullivan and Felton, *A Contested Election in California*, 21.

55. Ibid., 21–22.

56. Continuous-feed furnaces could be built to handle a variety of ore sizes, and these furnaces could equally well reduce rich or poor ore. Most mines mechanically crushed ore to the required sizes and may have done some rudimentary grading (often by sight), but for the most part the ore that came from the mine was simply run through the furnace.

57. This difficult issue is probably impossible to resolve; however, here is an important point to consider. Due to the geology of Mine Hill, particularly the irregular and thus unpredictable nature of the ore bodies, continuous prospecting was necessary for the continued life of the mine. The question, though, is, to what extent? Kenneth Rank–New Idria Mines Collection, 1867–1973, box 1559, folder 6, "E. R. Sampson to the President and Directors of the Quicksilver Mining Co., New York, July 7, 1869."

58. Almaguer, *Racial Fault Lines*. Lisbeth Haas, whose work is more spatial than Almaguer's, has studied identity construction in early California and argues that ethnic and national identities acquired their meanings "through the struggles between contending social groups over who had access to the land and to the rights of citizenship." Haas, *Conquests and Historical Identities in California*, 12.

59. Although the company owned all the property, the miners built and owned their houses on the hill. Some people who were not employees of the company or the family of an employee also lived on the hill, which makes the settlements at New Almaden different from many definitions of a company town.

60. For extensive photographs and history of the camps at New Almaden, see Jimmie Schneider, *Quicksilver*; and Lanyon and Bulmore, *Cinnabar Hills*. For insights into the Spanish-speaking community and the camps at New Almaden, see Pitti, "Quicksilver Community." For issues of race and ethnicity generally, especially concerning Native Americans, at the camps of New Almaden, see Coomes, "From Pooyi to the New Almaden Mercury Mine."

61. William H. Brewer, *Up and Down California in 1860–1864*.

62. Mary Hallock Foote was the wife of mining engineer Arthur Foote, who was employed as a surveyor at New Almaden. Her memoir is *A Victorian Gentlewoman in the Far West*. Wallace Stegner used Foote's writings as the basis for his novel *Angle of Repose*.

63. "A California Mining Camp," *Scribner's Monthly Magazine* 15, no. 4 (February 1878): 480–94, quoted in Foote, *New Almaden*.

64. Phyllis F. Butler, "New Almaden's Casa Grande." Halleck was also largely responsible for the construction of the Montgomery Block, the largest office building in San Francisco, in the mid-1850s. Halleck later served as President Lincoln's general-in-chief.

65. There are few records on the original development of Spanishtown; however, it is known that Spanishtown residents owned their houses and paid no rent to the company. But although Spanishtown residents had some autonomy, Barron, Forbes & Co. played an active role in the development of the town, providing money and materials to see that its workforce, imported to the mine at significant expense, was provided for.

66. The English, or Protestant, cemetery at New Almaden was in the Hacienda. Over time a second Catholic cemetery, the Hidalgo Cemetery, was established around the other side of Cemetery Hill from Spanishtown.

67. Philip Scranton, "Varieties of Paternalism." Scranton argued that there are many varieties of paternalism, each determined by the quality of the interrelations of the three overlapping spheres of paternalism: provision, protection, and control. Paternalism emerged as a term

in the early 1880s to describe a system whereby an authority supplies the needs and controls the actions of individuals and groups under its control.

68. Ibid.

69. Reasons for this lack of interest include that collecting land rent was not a traditional practice for them and that they were making so much money that the land rent was inconsequential.

70. Pitti, "Quicksilver Community," 125.

71. Workers could travel over the hills to get to San Jose, but the hills were rugged and the distances great. As would be expected, there are accounts of black markets and contraband goods.

72. Strikes also occurred in early January 1865 (four days); January 30, 1865 (one day); March 1866 (four days); and April 1866 (eleven days). For a detailed analysis of these events, see Pitti, "Quicksilver Community"; and Coomes, "From Pooyi to the New Almaden Mercury Mine."

73. Pitti, "Quicksilver Community," 120–32.

74. These newspaper accounts are along the lines of the one in the *Daily Alta California*, April 10, 1866: "A strike took place among the miners at New Almaden on the Hill, to-day. The strikers are said to be mostly Mexicans: they have taken possession of the mine and will allow no one to work. The grievances they complain of are the same as on a similar occasion last year: the rules of the company which compel them to lease or rent houses they have built on the company's land, and being compelled to purchase supplies only of dealers who pay the company a bonus for the privilege of selling, thus compelling them to pay higher prices than when peddlers were allowed to visit the mine."

75. The Mexican protestors, as argued by Pitti, connected their oppression at New Almaden with those colonial powers of Europe who at the time were attempting to take over Mexico; the protesters particularly connected their situation to the French invasion of Mexico in 1862.

76. New Almaden Mine Collection, 1845–1973. The French phrase can be translated as "that which is deferred is not past."

77. Pitti, "Quicksilver Community." Pitti makes these arguments in chap. 2.

78. The 1860 census data, unfortunately, are not reliable enough for a comparative chart. The mine was closed by injunction at the time of the 1860 census, and although many of the mine community were recorded in the census, the mine closure prohibits a direct comparison.

79. Pitti, "Quicksilver Community." Pitti makes this claim in chap. 2. Coomes, "From Pooyi to the New Almaden Mercury Mine." Coomes makes this claim in chap. 3.

80. *San Jose Mercury*, October 6, 1890.

"The Yard Gang," a photograph from the early twentieth century, shows a group of thirteen men and a dog at the New Idria Mine in California (Figure 5.1).[1] Together these men—the reduction yard workers—sorted and crushed the ore coming from the mine, loaded the ore into quicksilver furnaces, and then bottled the resulting mercury for market.[2] The photograph is most interesting for its careful composition, which highlights the many different races and ethnicities of the men who worked "the yard" at the mine. Whoever owned the postcard recognized the photographer's message, for on the back is handwritten: "Quite a cosmopolitan bunch. Spanish, German, Korean, Italian, Mexican, Italian [repeated in original] and American and Hindu." This postcard evokes a nostalgic theme of California, especially from gold rush days, of people coming from around the globe to reap riches and to forge a new life there. Here again, in the early twentieth century—sixty years after the gold rush—this photographer at New Idria celebrated this theme.[3] The "yard gang" was a remarkable group, marking the New Idria Mine, and by extension California, as home to people of many races and backgrounds. Due to this multiplicity California and New Idria were places where race relations were not simple oppositions, such as black-white; instead, they were complex and many-tiered.[4]

For the "cosmopolitan" group of men in the photograph, race and ethnicity were major factors in dictating their work and working conditions at the yard. For example, the 1910 manuscript census for New Idria lists eleven Koreans, all of whom worked on the quicksilver furnaces, the unhealthiest job at the mine and the job that was held historically by the lowest group in the racial hierarchy there.[5] Thirty years earlier, Chinese men had done all of the furnace work at the mine (Figure 5.2).[6] The hierarchies changed over time as new groups were inserted, or as groups left the industry. However, groups never changed their relative positions within the hierarchy.

Although the industry underwent change from its initial development in the 1840s and 1850s, it was change based

5

Race, Technology, and Work

DOI: 10.5876/9781607322436:c05

FIGURE 5.1 The Yard Gang, New Idria Quicksilver Mining Company, Idria, California, 1910s
This picture postcard shows a few of the men who worked the furnace yard at the New Idria Mine. (Courtesy of the California History Room, 1998-0442.)

on adaptation and modification of a model that remained fundamentally intact. The New Almaden model of Barron, Forbes & Co.—based in its British and Spanish imperial roots—allowed a small group of people to control the production, trade, and use of mercury in order to produce wealth for themselves. This model was maintained until circumstances spun out of their control in the early 1870s during the quicksilver boom. During the quicksilver boom and bust of the 1870s, there were four competing models of mercury mining: that of original mines such as the New Almaden Mine, that of the transitional Redington Mine, those of the new mines of the boom such as the Guadalupe Mine and the Sulphur Bank Mine, and that of the more informal boom districts. These four models give insight into the development of California through a detailed look at how social, cultural, industrial, and technological transformations were negotiated and contested in the 1870s.

Four Mine Models, One Racial Hierarchy

There were a number of similarities among the four mine models of the 1870s. In all of these models, mining, ore reduction, and sales were integrated and quicksilver

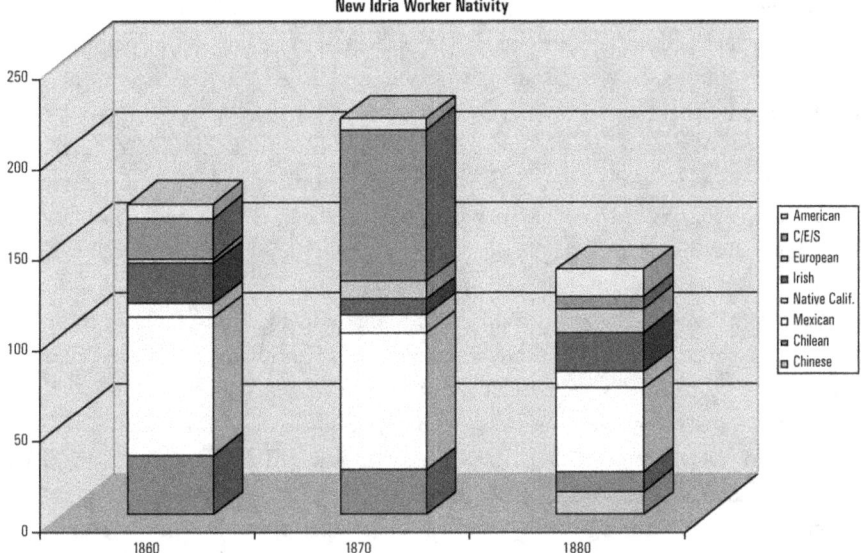

FIGURE 5.2 Nativity of Workers at the New Idria Mercury Mine, 1860, 1870, 1880
The New Idria Mine was controlled by Thomas Bell and his partners in the Bank Crowd over these years and was operated to enable their control of mercury markets. In 1860 the New Almaden Mine was closed by injunction, and production was up at New Idria; in 1870 the price of quicksilver was rising, and higher-cost Cornish miners were building extensive mine infrastructure; in 1880 Bell wanted modest but steady production from the mine and had a lower-cost workforce of Mexican miners in place. Determination of nativity is based on each worker's listed country of birth in the census. The abbreviation C/E/S stands for Cornish/English/Scottish. The category European is used for all countries in Europe not specifically called out. As an aid to reading the chart, the 1880 graph shows all eight groups. (Records of the Bureau of the Census, Manuscript Schedules of Decennial Population Censuses, 1860, 1870, and 1880, New Idria, California.)

production was the primary focus. Groups of people both lived and worked at each mine, and there were marked class distinctions at each mine among owners, managers, and workers. The major defining characteristic of each of these models was the racial and ethnic composition of their workforces. Although New Almaden (as we saw in chapter 4) had a racially complex workforce in which the association among different racial and ethnic groups and different jobs was normalized, at the new mines there was a simple line between white and Chinese: Chinese miners did all of the laboring jobs under a small group of white managers. During the boom years, the contrasts within the industry were stark: whereas Chinese men were ore pickers (a low-status job) at the New Almaden Mine, one of the original mines, Chinese men performed all the jobs at the Great Western Mine, one of the new mines, including being timbermen, a high-status job. At the Redington Mine, established between the old and new models in time, there were white, Protestant owners and an Irish, Catholic workforce. In the 1870s Mexican Catholics were hired at the mine, but there were never any Chinese employees. Unlike the workers at any other quicksilver mine in the state at the time, the Redington workers may have been unionized in some fashion. In the informal boom districts, white Anglo-Americans filed mine claims and often worked these claims themselves, occasionally hiring a small number of Mexican or Chinese laborers. Whites were the only group that could readily file mine claims in these districts, due to both state laws that governed who could file claims and miners' conventions, whose regulations governed local districts.

When a quicksilver mine was developed mattered because time determined the nature of the racial and ethnic relations of the workforces. The workers' nativity chart based on data from the 1880 manuscript census tells an important story (Figure 5.3). Original mines, the three on the left of the chart, first developed in the 1840s and 1850s, and by later decades had workforces with a complex racial and ethnic structure that closely mirrored the structure in place at New Almaden.[7] By comparison, the new mines, the three on the right of the chart, were all developed during the quicksilver boom years of the early and mid-1870s and had what was in practice a workforce with a simple dividing line between white and Chinese.[8] The Redington Mine was a transitional mine that from its development in the mid-1860s had a two-tiered workforce of all whites but with class position as key, with Irish Catholics as the laboring class and Protestants as the managing class.

In spite of significant technological changes in the mercury industry that standardized the tools of the industry by 1875, neither the organization of the workforce nor the nature of the working population became standardized across mines. Although all of the mines had shifted from intermittent furnaces—which had to be loaded, fired, cooled, and then unloaded—to furnaces that were self-feeding and were fired continuously, the work and the other tasks involved in mining and refining mercury were organized differently under the four different models of mines, and their workforces had different racial and ethnic compositions. Technology did not determine the organization of labor, therefore, although it did define the tasks that needed to be fulfilled. We must turn instead to the ethnic and racial composition of the workforce at each mine in order to better understand the organization of work. Records from the mines and other sources can be used to isolate a racialized hierarchy of job status for the four mine models that allows us to see how the intersection of technology, race, and ethnicity shaped the nature of labor at each mine.

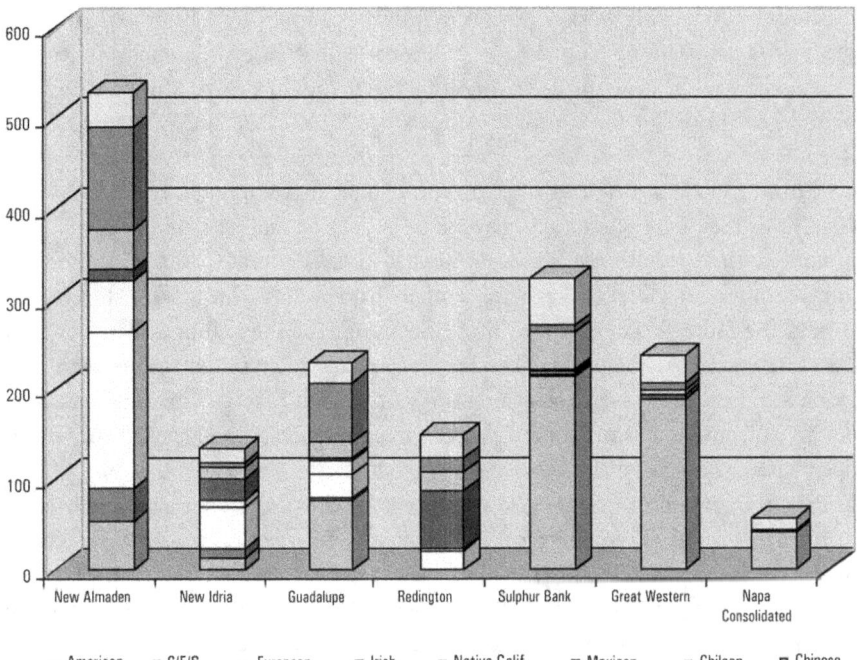

FIGURE 5.3 Nativity of California Mercury Mine Workers, 1880

The three original mines, shown on the left, had a complex racial and ethnic hierarchy. The three new mines, on the right, had a two-tiered workforce of white managers and Chinese workers. The Redington Mine was a transitional case in which the mine had a two-tiered workforce but all employees were white. Mexican workers were employed at the Redington Mine beginning only in the late 1870s, when the mine was struggling financially. As an aid to reading the chart, the graph on the left, New Almaden, shows all eight categories. (Data are from the Records of the Bureau of the Census, Manuscript Schedules of Decennial Population Censuses, 1880.)

Race and Work at the Original Mines

Before the quicksilver boom, the organization of work at the Guadalupe and New Idria Mines was similar to that at New Almaden. Mining companies—Anglo-American and Mexican at Guadalupe, and Mexican, Chilean, and Anglo-American at New Idria—had yardage and tribute contracts for underground work. Through the boom years, the racial composition and labor organization of both the Guadalupe and New Idria Mines became more complex, particularly with the introduction of Chinese laborers.

At the Guadalupe Mine, records indicate that managers experimented with various methods of organizing work. These were influenced by changes in ownership from the Santa Clara Mining Company of Baltimore, which held the mine from its first development in the 1850s until 1874 or early 1875, to the Guadalupe Mining Company, partially controlled by the Comstock silver kings James Flood and William O'Brien. Records show that in 1874, workers were paid by the day, and there were apparently no Chinese workers. The records under the Guadalupe Mining Company, in contrast, show that work was soon organized by a series of contracts held by companies of miners. These records also show payment to Chinese workers, through a monthly lump sum for forty-one to sixty-five workers paid to a "Chinese boss." From 1876 to 1881 the monthly pay to the Chinese boss averaged thirty-four dollars per man (most men were paid $1.25 a day, but some were paid only $1, whereas five men who worked at the furnaces were each paid $1.50). An average of seven dollars per man was deducted from the pay of Chinese workers, resulting in an average of twenty-seven dollars per man per month net.

The contract system appears to have lasted until an upheaval occurred at the mine in June 1880.[9] The company records show that 129 employees of the mine—Euro-American, Mexican, and Chinese—sued the company for wages due them (the suits were for pay from around June 1880 and were filed in November 1880). Sixty-one Euro-Americans sued for an average of $128, twenty-two Mexicans sued for an average of $85, and forty-six Chinese sued for an average of $63. By November 1880, most workers at the mine were paid by the day (a few workers doing the most skilled jobs were paid by the month), and the disruptions and legal suits of the summer seem to have been resolved.

In 1880, prior to the lawsuit later that year, the mine was working three shafts. Two of these three had companies of workers with European or American names. The third shaft, the one with the most workers, had about 60 percent Mexican workers and 35 percent with European or American surnames. All of the work sorting ore was done by Mexicans, often boys. The Chinese at the mine mostly worked tramming ore. Each of the shafts had a number of Chinese workers listed as "Car-men" at $1.25 a day. Chinese workers were also the "firemen" tending the furnace fires, and soot men who cleaned the furnaces. Unlike at New Almaden, at the Guadalupe Mine management employed Chinese workers underground, at the *planilla*, and at the reduction works. These Chinese workers held the lowest jobs (often support jobs), and they were paid less than other laboring groups, but they worked in the same spaces as the Euro-American whites and Mexicans, unlike at New Almaden.

Whereas Chinese men worked everywhere at the Guadalupe Mine, the Euro-American whites and Mexicans did not.[10] Underground at Guadalupe, miners worked on three levels: 6, 7, and 8. Only Euro-American white miners worked on levels 6 and 7. Level 8, the main mining level, had 16 Euro-American white miners, 45 Mexican miners, and 19 Chinese trammers. Maps of the mine show level 8 to be extensive, and it is probable that the Euro-Americans and Mexicans worked in separate areas of it. These facts from the company records show that, as at New Almaden, the mine workers were divided into racial and ethnic groups and that work space was racialized and largely segregated. At the planilla all of the workers were Mexican or Chinese, whereas at the reduction works most of the workers were Euro-American whites (who were supervisors) or Chinese, joined by only four Mexicans. At all points the Chinese workers were hired to do the most menial and most dangerous jobs, including tending the furnaces.[11]

At New Idria the workforce composition was similar to that of the other old mines, yet it is also evident how William Barron and Thomas Bell manipulated the workforce to suit their goal of increasing or decreasing production on demand.[12] The chart comparing the nativity of workers at New Idria for 1860, 1870, and 1880 shows significant changes to the workforce over these years (see Figure 5.2). In 1860 the workforce was mostly Mexican and Chilean, supplemented by Cornish, Irish, and American miners. Around 1870 the mine management added a large number of Cornish miners, brought in to develop the mine rapidly at a time when the New Almaden Mine's production was floundering. However, by 1880 Thomas Bell did not want much quicksilver from New Idria; he did not fund the mine development work that the Cornish miners did in the industry at this time, and the Cornish workers all but disappeared from the workforce. Replacing them as "whites" at the mine were Irish workers—the lowest-cost white workers—and Chinese workers were hired to work the quicksilver furnaces, further lowering labor costs while transferring the dangers of the work to racial and ethnic groups that were lower on the racial hierarchy.

Race and Work at the Redington Mine

Although the Redington Mine was part of the original combinations with New Almaden and New Idria, it was also very different in that for its first decade of operation (1863–73), it had a simple two-tiered workforce.[13] In 1870 the census records show that the mine had Anglo-American owners and managers and Irish Catholic miners and laborers. The organization of work at the mine was modeled on gold and silver mines—the type of mines with which the mine owners had close connections—in California and Nevada. Mine foremen for the Redington were hired from the gold and silver mines and returned to them when done at the mine.[14] Further strengthening this connection were the combination agreements, detailed in chapter 2, through which the Redington Mine had the right of exclusive sales of quicksilver to the gold and silver mines of California and Nevada.

The ore deposit at the Redington Mine was large, very regular, and near the surface, comprising the mass of a gentle hill. On the plus side, this meant that mine infrastructure did not need to be extensive or technologically advanced. Simple, inexpensive methods were adequate to extract the ore. In addition, because the ore body was regular and near the surface, there was little need for expensive exploratory work. However, the Redington ore was low-grade, and mining was complicated by problems with water and gaseous carbonic acid in the mine, mitigated by pumps and ventilation shafts.[15] By the late 1870s the Redington Mine had worked through most of its ore and must have been under enough financial pressure for its managers to have hired Mexicans and Chileans as part of the workforce. This bow to the rest of the industry would have at the same time distanced them from the gold and silver mines of California and Nevada that they had emulated up to this time.

A survey of the 1880 census for job titles shows that of the highest-status jobs five were held by Irishmen, fourteen by other Euro-American whites, and none by Mexicans. The Redington Company appears to have hired men by the month, their wages including room and board or, if they had a family, a company-owned cottage. The men worked the standard two 10-hour shifts. The census enumerator (census taker) for Knoxville used four standard job categories for men: "works in mine," "works on furnace," "laborer," and the proper name of (mostly) higher-status jobs, such as carpenter, teamster, or blacksmith. The only jobs held by Mexicans at Knoxville were in the mines. Irishmen, and a few other Euro-American whites, did most of the furnace work, and all the general laborers were white or Irish. The Redington Mine is the only quicksilver company that did not employ any Chinese workers, reinforcing the company's status as an outsider.[16] Their position on Chinese labor won them praise in local newspapers that was tempered by business realities as the mine failed in the late 1870s:

> Why the Redington cannot prosper as other well conducted mines in this portion of the State appear to, is more than we can comprehend, unless it be that the expenses of the company are greater than those of other quicksilver mining companies, caused (be it said to the Company's credit) by employing white men only, at fair wages.[17]

Race and Work at the New Mines

The new mines, of which the Great Western is the best documented, were opened during the quicksilver boom years of the mid-1870s.[18] These mines were generally owned by California capitalists and worked primarily by Chinese labor. As the 1880 nativity chart shows, each of the new mines had a two-tiered workforce of Chinese laborers supervised by a small number of whites (see Figure 5.3). These new mines reshaped the industry through less capital-intensive mine development in both mine and camp infrastructure that was possible in part because of the low status of Chinese workers in the prevailing racial hierarchy.[19]

The lower cost of Chinese labor, combined with the price boom in quicksilver markets, made numerous known mercury mine prospects candidates for major development in the early and mid-1870s.[20] Geologically, these new mines had substantial quantities of low-grade ore, accounting in part for why they had not been developed previously; the original mines, at least when first developed, had high-grade ore and had reduction technology designed to process smaller quantities of high-grade ore. After the Redington Mine pioneered the reduction of large quantities of low-grade ore in the late 1860s and early 1870s with its continuous-feed fine-ore furnaces, and after New Almaden marketed its Scott furnaces, new mines adopted this technology and prospered. But although the new mines all used similar reduction technology, they differed in working conditions for the miners and the technology that was necessary for mining the ore.

The Sulphur Bank Mine was unique among Californian mercury mines in that there were few shafts or tunnels or underground work at all—most of the ore was shoveled up from open cuts in the surface. The Napa Consolidated (Oat Hill) and the Great Western were extensive mines with relatively easily worked ore bodies near the surface. At Oat Hill, the ore bodies took the form of large lenses of sandstone permeated with cinnabar. Although shafts were dug to 400 feet on one vein and 750 feet on another, most of the ore was mined from stopes just beneath the surface. Similarly, the long dimension of the ore body at the Great Western ran parallel to the mountain ridge in which it was located.[21] The ore body (over 1,000 feet in length) was tapped by a series of tunnels dug at right angles to the ore (Figure 5.4). One of these tunnels was driven through the mountain and was used for removing the ore, and furnaces were built at each outlet for reducing the ore. Over time the mine was worked deeper by means of internal shafts (winzes) through which ore was brought up to the main tunnel. The mine did not have serious water problems, but it did require extensive timbering. The ore was richer as the miners worked down through the deposit; however, it generally averaged only 1 to 2 percent. Mostly, however, these new mines required no extraordinary technology or infrastructure, as did the deep shafts at New Almaden or the water problems at Guadalupe, combatted with an elaborate inclined shaft.

Although quicksilver mining companies had experimented with employing Chinese miners since the late 1860s, it was not until the quicksilver boom, and the

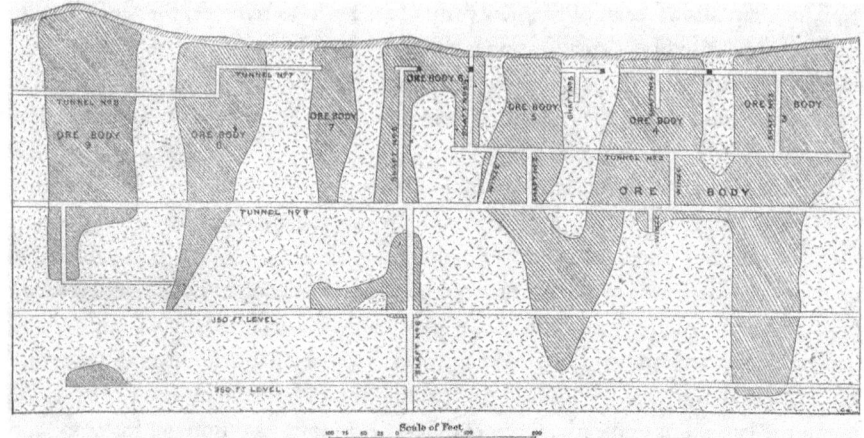

FIGURE 5.4 Longitudinal Section of the Great Western Quicksilver Mine
The ore at the Great Western Mine was in the contact between sandstone and serpentine. This section shows chimneys of ore and the extensive and regular mining methods used to work it. The richest ore was near the surface, and there was also substantial native (pure) mercury in the mine. Scale is shown. (Becker, *Geology of the Quicksilver Deposits of the Pacific Slope*, fig. 16.)

development of the new mines, that they were employed in large numbers.[22] The racialized workforces at the original mines had been shaped over many years and Chinese workers could not easily have been injected into these workforces without radically adjusting workforce composition and the corresponding organization of work. At the new mines, however, the owners and managers were able to invent new models of workforce composition and organizations of work. Chinese men formed a core of workers with valuable skills who could be employed by the new mines. Many Chinese men had years of mining experience in California; as the *Mining and Scientific Press* reported on February 2, 1875, "Over one hundred Chinese have left Columbia (a gold mining town in the Mother Lode) since Friday week—most of them for the new quicksilver mines."[23] In addition, many Chinese men had gained experience valuable in mining, especially drilling and blasting, by working on the railroads.

Although some of the Chinese men at the quicksilver mines were skilled miners, many if not most of the Chinese men hired by the new mines were simple laborers who learned the necessary skills on the job. These men happened to end up working at the quicksilver mines, but they could just as easily have ended up in agriculture in the central valley, or in a cigar factory in San Francisco. Many Chinese immigrants came to California as sojourners, interested in making money on "Gold Mountain," as California was called, in order to pay for a better life when they returned home.[24] Sojourners were ambitious and looking for social mobility. Most came from modest, but not poor, backgrounds. Often their travels to California were part of a family strategy by which young men traveled and parents, wives, and children remained behind. For the sojourners, home and family remained fixed in China despite their new lives abroad. These men were part of a social system of labor contracts that stretched from China across the Pacific. Upon arrival in San Francisco many of the men were in debt for their passage, and for the men in debt—and indeed for all Chinese men—it was nearly impossible not to be part of the web of controls exercised by their countrymen in California.

Chinese labor was organized vertically. At the top were the Chinese merchants of the Six Companies who controlled men who were indebted to them for their passage to California. Merchant contractors of the Six Companies made arrangements to supply gangs of Chinese laborers to whoever wanted them, and the quicksilver mines participated in this market. Making the labor deals were agents and interpreters who brokered specifics with mine representatives. At the mines the Chinese worked in labor gangs headed by a gang foreman who dealt with the mine operators directly and who served as the merchant's representative at the site. Despite the dominance of the Six Companies, the labor-contracting business was not monolithic. Rival agents, work-gang leaders, or groups of laborers made choices about selling their labor. A quote from the Calistoga newspaper in 1878 gives a glimpse of the dynamics at work:

> Some fifty or sixty Chinamen came down from the Great Western Quicksilver mine last Saturday afternoon and took the cars for San Francisco... These Celestials all left the mine on account of a reduction of wages. Their places have been taken by others willing to work for the amount offered by the mining company.[25]

At the Great Western, the community of 250 or so Chinese workers would have been assembled from a number of smaller groups.[26] According to author Helen Rocca Goss, who wrote of growing up at the Great Western Mine, there were two Chinese camps at the mine; the inhabitants were separated by region and tong affiliation.[27] Correspondingly, men were separated in their work by clan groups with the idea of preventing or reducing friction among the groups (much as the Cornish, Irish, and Mexican workers were segregated at other mines). Within the Chinese population there were both better-paid, highly skilled workers and low-paid, highly expendable laborers. At the Great Western, one Chinese man, Ah Shee, was reportedly the head timberman, a high-status job. Another man, Ah Cat, cleaned the quicksilver furnaces, a low-status job, and experienced acute salivation.[28] Goss wrote that on payday the China bosses came to the house and worked out with Superintendent Rocca the amount they were to be paid. These men would then have had a number of demands on the money received, including payment to the Six Companies labor broker, payment to the Chinese merchant at the mine, and the payment to the individual workers based on their rank and position in the group.

The working conditions of the Chinese miners could be horrendous; the Sulphur Bank Mine had the worst physical conditions of the new mines.[29] In the 1870s the mine was worked by open cuts with little regard to system, resulting in a labyrinth of deep open pits and trenches where miners shoveled up the ore, following the small but rich veins that were present nearly everywhere. Judging by geologist George F. Becker's account in 1882, working in the open cuts was horrible: "Work in these cuts is so trying that few white miners have ever accepted employment in them a second day and almost all the labor is performed by Chinamen."[30] Another geologist wrote of the conditions: "The odor of sulphur dioxide is very strong so that it was difficult to breathe while placing a thermometer in one of the openings to observe the temperature . . . There is a sound as of a roaring furnace from below. How anyone could breathe in such an atmosphere, much less work, is a matter for wonderment."[31] Despite these difficulties, the plentiful and easily accessed ore resulted in both large production totals (over 70,000 flasks) for the years 1875–82 and the lowest cost of production of any mine in the state, giving the majority owners, Parrott & Company, significant power in the quicksilver industry at the time. Although conditions at other mines were better, they were still often quite unpleasant. For example, at the Oat Hill Mine, the ore released unusual quantities of ammonia. Generally reduction works produced a strong, often unbearable, odor of sulphur, but at Oat Hill, one commentator wrote, "ammoniacal fumes are virulent, and an operator is not able to endure them but a short time."[32]

The biggest difference between the original mines and the new mines was that the Chinese miners—as compared with the other laboring groups in the industry, such as the Mexicans, Irish, or Cornish—could be treated much more poorly. Concerns on the part of management for working conditions and worker health were minimal. The low status of the Chinese workers meant that they were simply a commodity, a supply of labor, purchased in groups that could be hired and fired at will. The labor contractors of the Six Companies or similar institutions were subcontractors who delivered the purchased commodity: mine labor. For the mine managers it meant little who the individual was doing the work; what mattered was that a body was present to do the work at the agreed-upon price.

Whites at the new mines generally made up 20 percent to 25 percent of the total workers, and they enjoyed being the privileged group at the mines. These men did the management jobs in the offices and supervised the work of the Chinese men above-

	New Almaden	Guadalupe	Redington	Great Western
Managers	American/European	American/European	American/European	American/European
Engineers	American/European	American/European	American/European	American/European
Timbermen	Cornish/Mexican	European/Mexican/American	Irish	Chinese
Miners	Cornish/Mexican	European/Mexican/American	Irish	Chinese
Muckers	Swiss	European/Mexican/American	Irish/Mexican	Chinese
Sorters	Mexican Youth	Chinese/Mexican Youth	Irish/Mexican	Chinese
General Laborers	Mexican	Chinese	Irish/Mexican	Chinese
Ore Pickers	Chinese			

FIGURE 5.5 Racial Hierarchy of Mine Labor at Selected Quicksilver Mines
This chart shows the dominant ethnicity of workers doing specific jobs at various mines. Managers are the highest-status job, and ore pickers the lowest. Managers and engineers excluded, jobs were not tied to a particular racial and ethnic group uniformly across the industry. Jobs were racialized only in respect to the dominant racial hierarchy as it needed to be maintained at a particular mine. New Almaden and Guadalupe are original mines, and Great Western is a new mine. (Based on answers to questions concerning occupation and country of origin in Records of the Bureau of the Census, Manuscript Schedules of Decennial Population Censuses, 1870 and 1880.)

and belowground. Whites would have worked in the same spaces with the Chinese, and in some cases probably learned some basic language skills in order to communicate directions.[33] Unlike laboring groups at the old mines—where Mexican, Irish, or Cornish miners did a variety of jobs—no one but the Chinese men actually did the brunt of the dirty work of mining and reducing quicksilver. The new mines offered the white workers who were employed at them an undeniably privileged status.

In comparing the original and new mines, and the variations within each group, what mattered most in the organization of work at the quicksilver mines was maintenance of the racial hierarchy. The job a man of a particular racial and ethnic group did at a mine mattered only in relation to which groups did which jobs at that particular mine. A job was not a Chinese job or a Cornish job industrywide.[34] Figure 5.5 is a survey from census records of race and ethnicity across job types at a sample of mines. Across the industry, members of each racial and ethnic group did many jobs at the mines and reduction plants.

Race and Work in the Informal Boom Districts

The informal boom districts—the fourth competing model of quicksilver mining during the 1870s—did not participate in the racialized hierarchy of labor in the other mines, because they were largely worked by individuals. Although only modest quantities of mercury were produced in the boom districts, at the height of the boom they caught the public imagination and were prominently featured in newspapers throughout the state and the West. For the first and only time in the history of the U.S. mercury mining industry, quicksilver mining was open to the small-time investor and prospector. Men and women of modest means could dream of becoming rich by way of a mercury mine. However, this dream was open only to the men high on the racial hierarchy—Euro-American whites and occasionally Mexicans—who could legally claim mines and own property.

The two major boom districts north of San Francisco Bay were the Pine Flat District in Sonoma County (see Figure 3.7) and the Sulphur Creek District, straddling Lake and Colusa Counties (see Figure 3.9).[35] Locals prospected these districts either for themselves or in the employ of a local group of investors. Only white men could file claims to develop mines in these districts. The Pine Flat District of Sonoma County was the most frenetic of the boom districts during the quicksilver rush. The district had first been prospected and claimed in the early sixties during the closure of the New Almaden Mine; however, there was no recorded production in the district until 1873, when the quicksilver price boom pulled prospectors back into the area. The district was located in the mountains near the Lake County line in what is known as the Geysers, a thermally active zone that was developed commercially on a small scale for sightseeing and as spas. The prospects in the area were the northwestern part of the Mayacamas District, which stretched from the Aetna Mine at Aetna Springs in a line almost to Cloverdale. As the price of quicksilver skyrocketed, the inhabitants of Sonoma Valley looked to the mountains of the Geysers area, remembered the earlier boom over ten years before, and relocated the claims. As the *Sonoma Democrat* printed on November 1, 1873: "We have no desire to create a mining excitement [but] ... during the past summer, in the northeastern corner of Sonoma County, almost at our doors, there has been partially developed the richest and most extensive cinnabar deposit in the world."[36] As an editorial in the *Russian River Flag* stated sarcastically: "A man here that don't own a quicksilver mine will soon be regarded as an old fogy. Arise, then denizens of Healdsburg who are still mineless, and hike to the mountains! Stake your claims, incorporate companies, issue stock ad infinitum!"[37]

The task for prospectors in the boom districts was proving their mine in order to sell it. The mining and reduction technology they used was quick, cheap, and largely for promotion, with enough underground work to extract ore and simple retort furnaces to produce mercury on a small scale. Some miners with available capital engaged in development work in their mines, and perhaps the construction of a Scott or Knox fine-ore furnace. One compelling source on the quicksilver rush and the Sulphur Creek District is a daybook kept by Nathaniel Disturnell, a quicksilver boom miner and former San Francisco bookkeeper, from 1874 to 1876.[38] Nathaniel and his brother Richard (referred to in the daybook as Dick) together owned a number of claims in the Sulphur Creek District. These two men, midlevel operators, developed their prospects with their own limited resources.[39] Because they owned claims, at least one by all accounts quite promising, they had some freedom to make choices: they could work when they wanted (which was most of the time, even Christmas Day), they could go to town when they needed to, and they could deal for supplies in a limited way. As owners they participated in miners' meetings when they were held, particularly early on when groups formed to prevent "claim jumping."[40] As their fortunes rose the Disturnells were, not surprisingly, happy to shed old relationships for new ones offering them more financial benefit. The Disturnells had some success at Sulphur Creek, and over the two and a half years chronicled in the daybook they staked out claims, sold off one promising claim, and then worked hard to develop another claim, eventually having enough success to hire a few men to work for them. In August 1874 the brothers decided to sell their partial ownership of the Empire Claim, giving them capital to develop their other claims. By January 1875 they had sunk two shafts, extracted and stored a quantity of ore, and built a rudimentary stone-and-brick retort furnace that they used by filling flasks with rich ore and then placing the sealed flasks in the furnace. This furnace relieved them from having to pay to pack ore to nearby mines, and to pay the other mines a cut from their production. By May 1875 the brothers were producing quicksilver steadily, a few flasks a week at roughly fifty dollars a flask, and they arranged to ship the flasks to San Francisco via the next-door Abbott Mine. About this time they started hiring men to work for them, at first one man but by 1876 four or five men. By the end of 1876 their production had increased significantly, but unfortunately for them the price boom was over.

As was the case for other miners in the boom districts, the Disturnell brothers developed their prospects with their own resources, using their own labor to get started and then selling off one of their prospects for capital to use developing another prospect. The Disturnells may have been a modest success story of small-time prospectors developing their own mine and seeing it prosper. They were fortunate to have prospected a good site, and despite nonexistent production figures on the Disturnell claims themselves, eventually their claims became part of the Abbott Mine, the largest producing mine in the district, with 30,845 flasks to 1917.[41]

Maintaining the Racial Hierarchy

The most important distinction among the many mercury mines operating in the 1870s is the organization of work and laborers at each. Despite using largely the same technology, these mines demonstrate that there were many possible variations in how to run a mercury mine in the 1870s. Each of these mines had very different methods of organizing work, and different workforces. Race and ethnicity go far in explaining why these four models were so different. There was an established racial hierarchy in California, and these mines slotted themselves into this hierarchy based on their own circumstances. Although which group did what job differed from mine to mine, the racial hierarchy remained consistent across mines, such that the hierarchical relationship among Irish, Mexican, and Chinese workers, for example, always remained the same. Each of the mines accessed the same racial hierarchy at the lowest levels possible given their particular histories.

The technological change of the 1870s allowed fine ore to be processed easily, but this change is less important than the fact that the Chinese miners who worked through these changes at the new mines were paid significantly less than the Irish workers at the Redington Mine. Although the nature of the work changed during the boom years, both Chinese and Irish laborers worked through these transitions at the respective mines. The work required skilled as well as unskilled labor, and all racial and ethnic groups did both kinds of jobs. In addition, there was stratification within the laboring groups, in which some workers were long-term, highly skilled overseers, whereas others were interchangeable and often temporary gang labor. What differed was that the Irish workers were not as cheap as Chinese workers, and they were not as expendable.

There were some strong similarities among all the mines. Most of the original and new mines worked on some form of contract labor; mine managers at both the original and new mines contracted with (or for) workers on performance-based contracts. At an original mine such as New Almaden, these contracts were bid on by workers who formed their own companies. At new mines the contracts were negotiated between the mine management and Chinese labor brokers; workers as individuals or in groups had little say in the process. The miner/owners in the boom districts did not work on contract, but neither did they work for a daily wage; they worked for themselves. Because workers in the quicksilver industry did not work for a daily wage (except at the Redington Mine), the industry was different from any other corporate mining enterprise in the state, where most workers were paid a daily wage.

The competition among these models of quicksilver production was made possible by the large commodity value of mercury in the 1870s and early 1880s. While at each mine cinnabar was taken from the earth and reduced, the means and methods varied. The original mines had the richest ore and the most extensive ore bodies. By 1880 the easy ore had been extracted from these mines, but only New Almaden and Guadalupe (mines on closely related deposits) required men skilled at making deep shafts and huge pumps in order to continue. The other original mines, along with all of the newer mines, did not require mining high technology. For reducing the ore all mines required the continuous fine-ore furnaces because the rich ore at the original mines was exhausted and the new mines had mostly low-grade ore. Once a mine had one of these furnaces properly adjusted for its particular ore, it was competitive with other mines as far as reduction. The original mines were more productive: productivity figures, based on the number of flasks of mercury produced at the mine each year, divided by the number of workers, were 44 flasks per worker at New Almaden, 27 at the Great Western, and 33 at Sulphur Bank.[42] However, the much lower cost of labor farther down the racial and ethnic hierarchy, particularly Chinese labor, meant that the profit margins were higher for the newer mines in spite of their lower productivity. In 1880, each flask of mercury cost $22 in labor at New Almaden, whereas at the Great Western this cost was $19, and at Sulphur Bank it was $13.[43]

Notes

1. The postcard is in the California State Library and is filed under "Mines & Mining: Quicksilver: New Idria Mines," 1998-0442.

2. The 1910 manuscript census for the New Idria Mine lists many more than thirteen men working in the "yard," but these other men add no additional diversity in terms of country of origin to the group of men in the photograph. Records of the Bureau of the Census, Manuscript Schedules of Decennial Population Censuses, 1910, San Benito County, California, Panoche Township. (Page numbers and enumerator are illegible.)

3. For more on this theme of multiethnic and multiracial societies in the gold rush, see Rawls and Bean, *California*; J. S. Holliday, *The World Rushed In*; Holliday, *Rush for Riches*; Susan Lee Johnson, *Roaring Camp*.

4. Tomas Almaguer and Ronald Takaki, as well as New Western historians such as Patricia Nelson Limerick, have argued this. Almaguer, *Racial Fault Lines*; Takaki, *Strangers from a Different Shore*; Takaki, *Iron Cages*; Patricia Nelson Limerick, "Disorientation and Reorientation."

5. In 1910 there were three active mines in the New Idria Group: the New Idria, the San Carlos, and the Aurora. Koreans worked the furnaces at all three of these mines. Records of the Bureau of the Census, Manuscript Schedules of Decennial Population Censuses, 1910, Panoche Township, San Benito County, California. (Page numbers and enumerator are illegible.)

6. Records of the Bureau of the Census, Manuscript Schedules of Decennial Population Censuses, 1880, Panoche and Vallecitos, Fresno County, California, 2–11, Chas. Moore, enumerator.

7. At the original mines, for the most part, a small number of American managers oversaw a laboring workforce composed of Cornish, Irish, and Mexican workers. The most notable variations are that the Guadalupe Mine had a sizable percentage of Chinese workers, whereas

the Redington Mine had no Chinese workers. Interestingly the largest laboring group at the Redington Mine was the Irish, complemented by Cornish and later by Mexican workers.

8. There is no census data for the boom districts—they appeared and disappeared between the census years of 1860 and 1870 for the first boom during the closure of New Almaden, and appeared and disappeared between the census years of 1870 and 1880 for the boom in the mid-1870s.

9. The court battles over Chinese exclusion in early 1880 may have been used as an opportunity for reorganizing work at the Guadalupe Mine.

10. Of the 223 employees at the mine, 51 percent worked underground, 31 percent worked aboveground, and 17 percent worked at the reduction works. Overall the Euro-American workers averaged $2.55 a day, the Mexican workers $1.79, and the Chinese workers $1.20.

11. Chinese workers also did the furnace work at the New Idria Mine, according to the 1880 manuscript census. Except for the census data, comparable information for the organization of work and the workers unfortunately does not exist for the New Idria Mine.

12. The New Idria Mine was originally run on contracts much as at New Almaden. See New Idria Quicksilver Mining Company Collection #2 (1931), box 1549, folder 9, Clarence Hillhouse, "New Idria Quicksilver and Letters 1858–1859." Page 15 of this collection details a contract between Daniel Gibb, an owner and manager, and J. T. Meyers for a tunnel at twenty-one dollars per yard. Page 24 details a letter written on October 4, 1858, that reads, "Yesterday we reorganized everything at the mine. All but one set of men are working by contract . . . We abolished the provision system there altogether and now two restaurants are started. It was not done too soon, as they were eating us out of house and home." In these letters the reduction plant area at New Idria is referred to as the "Hacienda." These letters date from the period when Barron & Co. gained control of New Idria.

13. The major stockholders in the Redington Mine Company were John Redington, George Cornwall, and Horatio Livermore, whose brother, Charles, was superintendent. Redington was owner of the Redington Drug Company of San Francisco, a major concern in the 1860s that gave the mine company strong financial support and sales outlets throughout California. For a history of the mine, see Donald O. Haus, "The Knoxville-Redington Mine, 1860 to 1882."

14. An inventory of mine buildings at Knoxville included a "Union Hall." If there was a miners' union at Knoxville, this would have been unique in the history of the California industry in the nineteenth century. Randol, *Papers Relating to Quicksilver Mining, Ca. 1849–1894*, report on the Redington Mine, 1876.

15. Workers at the Redington Mine suffered from salivation with the same frequency as workers at any other mercury mine. From the front page of the *Mining and Scientific Press* of March 16, 1872: "Quicksilver Fumes—We hear much complaint with regard to the deleterious influence of quicksilver fumes both from retorting the precious metals and from furnace operations in treating the ores of cinnabar at the quicksilver mines. Mr. William Kringle, who has been for some time employed at the Redington Mines, at Knoxville, Napa county, informs us that many are severely troubled in that way at that mine." In describing the process of cleaning the condensers at Redington, mining professor Egleston wrote: "On account of

the dust arising from the falling soot, it is necessary, when this work is done, for the men to wear a wet sponge over the mouth and nose, which is covered with a thin cloth tied behind the head. At this work the men relieve each other every fifteen minutes." Egleston, "Notes on the Treatment of Mercury in North California," 284.

16. As Redington Mine superintendent Charles Livermore put it in 1881, "Owing to the present low prices of quicksilver, the number of men employed now, and during the past year, is very small, being a total of about fifty, all of which are white men, there never having been a Chinaman employed about the premises by the present superintendent." Lyman Palmer, *History of Napa and Lake Counties*, 164. Livermore was making this statement at a time when Chinese labor was a major issue in California and the nation. It was a position that put him (and the Redington Mine) at odds with many rich and powerful men in California and in the quicksilver industry. See Andrew S. Johnston, "Quicksilver Landscapes."

17. "Quicksilver," *Independent Calistogan*, January 21, 1880.

18. Other new mines that had a reliance on Chinese labor include the Great Eastern, Socrates, and a few other mines in the Pine Flat area of Sonoma County; Saint John's Mine in Solano County; and the Oakville, Bella Union, and Aetna Mines in Napa County. The two other large mines developed in the quicksilver boom years, Altoona in Trinity County and Oceanic in San Luis Obispo County, were not active during the 1870 and 1880 census years.

19. Another reason is the continuous furnace reduction technology that made it profitable to process the ore at these new mines.

20. Many of the mines that were developed during the quicksilver boom, and all of the districts, had already been prospected and identified before the boom.

21. For an early report on the Great Western Mine, see *Mining and Scientific Press*, September 6, 1873, 152. For a personal story of the life of Superintendent Rocca and his family at the mine, see Helen Rocca Goss, *The Life and Death of a Quicksilver Mine*; and Helen Rocca Goss, *Gold and Cinnabar*.

22. For a detailed look at Chinese miners in a variety of contexts in the West, see Liping Zhu, "No Need to Rush."

23. *Mining and Scientific Press*, February 2, 1875.

24. Thomas Chen, *Chinese San Francisco*.

25. *Independent Calistogan*, February 27, 1878.

26. During the 1870s and early 1880s, the Great Western averaged 200 Chinese workers and about 50 European and American men; both of these groups were paid by the day (or by the month in a few cases).

27. Goss, *The Life and Death of a Quicksilver Mine*; Goss, *Gold and Cinnabar*.

28. Goss, *The Life and Death of a Quicksilver Mine*, 69.

29. The Sulphur Bank Mine site on Clear Lake in Lake County was first mined for borax in 1857, and by 1867 the California Borax Company used Chinese labor to produce both borax and sulphur from the site. By the quicksilver price boom of the early 1870s, the company decided to mine the site primarily for the cinnabar that was also present, although some sulphur continued to be produced. The Sulphur Bank Mine might have been the only mine

in the state where active deposition of cinnabar was occurring during mining operations, brought by the scalding sulphurous waters pushing up through and saturating the various strata of rock. Geologists, however, disagree on the theory of active deposition. See Donald L. Everhart, "Quicksilver Deposits at the Sulphur Bank Mine, Lake County, California."

30. Becker, *Geology of the Quicksilver Deposits of the Pacific Slope*, 254.

31. Walter Bradley, *Quicksilver Resources of California*. The *Independent Calistogan* described the underground conditions on July 20, 1881: "Gas is a source of great annoyance to the workmen, nauseating them, and making necessary frequent changes on each shift . . . the men changing off every few minutes." Another description of the Sulphur Bank Mine comments on one small shaft, twelve feet deep, that "had to be abandoned on account of the peculiar gas occasionally occurring in quicksilver mines, especially attacking the eyes of the men, blinding them temporarily with intense suffering." Aubury, *The Quicksilver Resources of California*, 65.

32. Palmer, *History of Napa and Lake Counties*, 171.

33. Goss relates that her father spoke a pigeon Chinese. Goss, *The Life and Death of a Quicksilver Mine*.

34. For a history of the Chinese workers at the new mines and their involvement with the anti-Chinese legislation of the New California Constitution, see Andrew S. Johnston, "Quicksilver Landscapes: Mercury Mining and Chinese Labor in Northern California to 1880."

35. Other boom districts included the Altoona and Integral Mine area in Trinity County, and the Cinnabar District near the Oregon border. South of San Francisco Bay there were numerous boom districts, the most prominent being Cambria in San Luis Obispo County. For complete information on these districts, see Bradley, *Quicksilver Resources of California Bulletin No. 78*.

36. *Sonoma Democrat*, November 1, 1873, as cited in Joseph Daniel Pelanconi, "Quicksilver Rush of Sonoma County, 1873–1875," 13.

37. *Russian River Flag*, February 11, 1875, as cited in ibid., 14–15.

38. Nathaniel F. Disturnell is listed in the 1880 census as thirty-nine years old and born in New York. His brother Richard O. Disturnell was forty-four and was also born in New York. Both men were single and in 1880 were boarders with Josiah Gentry and his wife, Mary, in the Lower Lake Precinct, District 49, Lake County. The Gentrys were farmers in the area of Sulphur Creek; the daybook notes that Nathaniel bought milk and eggs from them. Disturnell, "Diary Relating to Mercury Mining." Before the quicksilver rush, Nathaniel worked as a bookkeeper for the Merchant's Mutual Marine Insurance Co. He is listed as working there in the 1864 *Towne and Bacon San Francisco City Directory* and the 1872–73 *Cook and Miller City of Oakland Directory*. Richard is listed in the 1864 directory as a conductor for the Central Railroad Company.

39. From Disturnell's daybook, the district had two developed mines with groups of paid employees and expensive furnaces—the Abbott and the Buckeye—numerous small-time prospect owners such as themselves; a few businessmen running a store, hotel, and probably a saloon; and men who were trying to sell their labor, including whites, Mexicans, and

Chinese. There were also farmers and ranchers in the general area. Disturnell, "Diary Relating to Mercury Mining."

40. In June 1874 a law was passed "extending, until the first of January next, the time for expenditure of labor on mining claims located previous to May 10, 1872." *Napa Register*, June 13, 1874, as quoted in Joseph Daniel Pelanconi, "Quicksilver Rush of Sonoma County, 1873–1875," 25. Although Pelanconi strongly argued that this law caused a great deal of litigation concerning the mines in the Pine Flat District, it is also possible that this law raised issues of ownership at Sulphur Creek.

41. Bradley, *Quicksilver Resources of California Bulletin No. 78*, 53, states that the Abbott Mine "includes also, the old Disturnell."

42. These figures give only a rough estimate, based on published production figures for the mine and census records.

43. These figures are useful only for rough estimates, and are based on assumed average wages of $2.55 for whites, $2.20 for Irish, $1.80 for Mexicans, and $1.20 for Chinese. The same analysis of the Redington Mine and the Guadalupe Mine results in wildly unrealistic figures because of the radical changes going on at these mines in 1880. Of the seven mines analyzed here, the Redington and Guadalupe are the two that collapsed in the early 1880s.

Photographs showing a nineteenth-century California mercury mine community are rare. Figure 6.1, a photograph taken at the Great Western Mine in Lake County, California, shows members of that mine community in 1879.[1] This photo is of a type common to factories or company towns, in which workers, managers, and sometimes their families posed as a group in front of where they worked. Here members of the mine community pose on a hillside at the mine, with an ore chute to the right and the hillside denuded of vegetation, perhaps killed by mercury. Roughly 140 Chinese workers are grouped together in the back (upper) part of the photograph, standing and crouching on the hillside above a path cutting horizontally through the center of the image. Below the path are about twenty-five white men, four white women, and ten white children. In this photograph there is no intermixing of white and Chinese. The path through the center of the image is a demarcation line: above the line and in the background, Chinese, and below the line and in the foreground, white.[2] A few of the Chinese men are posing as individuals—most, however, are part of a mass bunched together behind the whites. The whites, in contrast, are identifiable as individuals, leaning on a tree, sitting on a stump, goofing on a wagon, or striking a pose in a top hat, such as the mine manager, Andrew Rocca, at middle right, probably joined by his wife sitting to his left on the ore chute. How the bodies of the people in the photograph are organized in space in this photograph is evidence of the racial organization at the mine. The Chinese are massed together at the rear in a more marginal position. The whites take up more space; the superintendent takes up the most of all. Notably, all of the women and children are white. White employees of the mine could have wives and children living with them; the Chinese employees could not. In the 1880 census for the Great Western Mine, there were 189 Chinese men, 46 white men, 25 white women, and 25 white children listed as living at the mine.[3] The racial organization displayed in the photograph reflects the landscapes of the camps.

6

Race, Family, and Camp Life

DOI: 10.5876/9781607322436:c06

FIGURE 6.1 The Great Western Mine Community, 1879
Photographs showing a mercury mine community are rare. Here Chinese workers are the majority, joined by white men and white women with a few white children. (Helen Rocca Goss, *The Life and Death of a Quicksilver Mine*.)

If the quicksilver mines—the shafts, tunnels, *planillas*, and furnace yards—were landscapes of production, then the quicksilver camps—the houses, outdoor areas, group structures, roads, and camps—were landscapes of reproduction. The camps were reproduction spaces for all of the people in the mine community, including men, women, and children. It was where the miners, laborers, managers, wives, and children ate, slept, and otherwise refueled their bodies and minds in order to work another day. The labor of reproduction, such as providing food and keeping living spaces clean, was done by women, children, and the elderly, as well as by the mine employees and providers of services, including shopkeepers, cooks, launderers, and water and wood haulers.

The landscapes of quicksilver mining camp life were fundamentally affected by struggles over race formation connected to the capitalist transformations going on in California.[4] Based on the composition of the workforce, a quicksilver camp could have definable white, Chinese, Mexican, Chilean, Cornish, or Irish landscapes, and the struggles among these groups had physical forms. The camps are best understood as sets of distinct racialized landscapes, tied to specific overlapping racial and ethnic groups; the quicksilver camps were the sum of all these racialized landscapes. The racialized camp landscapes had a physical component, separate areas inhabited by different groups, but the ways each group imagined the spaces of others—and the contrast between these imaginings and how the members of each group saw themselves—are equally important. In addition, people living at the camps interacted with those in other groups and sometimes experienced the spaces of others. These experiences in turn shaped and reimagined their landscapes.[5]

Quicksilver Camps in California

The camps associated with the four competing mine models during the quicksilver boom were even more structured by race than their work landscapes were. The original mines, the new mines, the Redington Mine, and the boom districts were racialized in different ways. The original mines continued to support a complexly racialized paternalism, whereby employees of the mine were treated according to their race and ethnicity and where they ranked on the racial hierarchy. This structure was reflected by their complex camp landscape. At the new mines there was a simple, two-tiered hierarchy of white and Chinese. The Chinese workers entering the quicksilver workforce in the early 1870s radically influenced the form of the new quicksilver camps as they had the landscapes of work. But the Chinese miners impacted the old and new mines in different ways. Common between the original and new mines, however, was that the Chinese lived more poorly than any other group. The Redington Mine also had a simple two-tiered hierarchy, but without Chinese workers. Although the camps for all of these models were company towns organized by owners, the camps in the boom districts were less formally organized, but still reflected the same racial hierarchies that underlay all of the mercury mining camps.

The Original Mines and Landscape Persistence

The original mines—New Almaden, New Idria, and Guadalupe—generally followed a household-based hacienda model with a complex hierarchy of races and ethnicities. When, in the early 1870s, Chinese miners entered the industry, each of the original mines hired and attempted to integrate these men into its workforce. Although racialized paternalism remained in place for the previously existing racial and ethnic groups of the mine workforce, the Chinese miners—at the bottom of the racial hierarchy—constituted a completely different system, a system that was not paternal in that although a strong aspect of control was enforced by the mine managers over their lives, there was little in the way of provision or protection. The Chinese men came to the mine as contract laborers; mine managers simply paid a "China Boss" for the amount of work done by the group. Much of the responsibility for their provision and protection was passed off to the labor contractors, which contributed to the enormous contrast between the spaces of camp life for Chinese workers and those for all other workers.

The Guadalupe Mine, for which there are a large number of visual and documentary sources, epitomizes the camps of the original mines. Its primary camp, Guadalupe Village, was located in a small and steep valley at the northwestern end of the ridge that contained the New Almaden Mine (Figure 6.2). The village was built essentially on top of the mine and then during the boom of the 1870s expanded up and down the valley. The camp needed to accommodate Euro-American, Latino, and Chinese workers: in 1880 the mine workforce was composed of 47 percent white workers, 19 percent Spanish-speaking workers, and 34 percent Chinese workers.[6] The formal village, however, did not accommodate all of the workers. It consisted of a central area with the store, boardinghouse, offices, and houses, and at least four separate groups of about six to eighteen company-built houses elsewhere in the valley. The Chinese miners did not live in company-built houses; instead, they built their own shanty villages in a company-assigned area that is not designated on the map. They would have been within walking distance of the sites of work but largely out of sight for the rest of the population.[7] The other racial and ethnic groups lived in the houses shown on the map, and the number of houses on the map (drawn in the early 1880s) corresponds well with the number of housing units recorded in the 1880 census. It is not known who lived in which of the houses or in which of the housing areas, but the census does give important clues as to the housing arrangements.[8]

Company-built cabins were the primary housing type and housed either a nuclear family or a small number of miners. These miners in all but one case lived strictly with members of their own racial and ethnic groups, that is, in a house of whites or in a Spanish-speaking house. One large boardinghouse housed thirty-six whites (two were women), and one family-run boardinghouse had seven white boarders. There were no Spanish boardinghouses at the mine; single Spanish-speaking miners lived as boarders with Spanish-speaking nuclear families, always in numbers of one to three boarders per house. Whereas this arrangement was common for the Spanish-speaking families, only four white families had boarders.

The non-Chinese population pyramid from the census information on the mine shows a community that more closely resembles an established village than it does a temporary mining camp (Figure 6.3). The pyramid shows that while definitely heavy with males of working age, there were many women, a number of elderly, and many

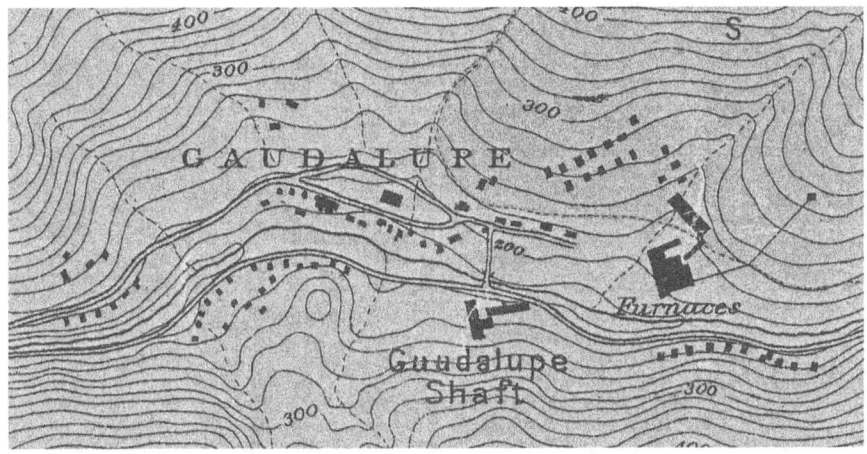

FIGURE 6.2 Plan of the Guadalupe Mine, 1880s (north is to the left)
New Almaden was a few miles off the map to the top right; the town of Mountain View is off the map to the left. The Guadalupe River runs through the middle of the valley, from right to left. At full development, as shown in this plan, the camp had distinct housing areas with about sixty houses, some of which are detailed in the accompanying photos. The superintendent's house is to the extreme left. The Chinese camp is not shown. (California State Mining Bureau map.)

children.[9] A second pyramid shows Chinese workers versus non-Chinese workers (Figure 6.4). The most important distinction is that as a whole, the Chinese workers were younger. The Chinese miners were younger, cheaper, and did not have families at the mines who needed to be supported. In addition, there were no Chinese who were not workers.

From 1881 to 1883 a series of photographs of the Guadalupe Mine were taken showing the mine workings and many aspects of Guadalupe Village, although none show the Chinese camp.[10] The photographs show a remarkably orderly and picturesque camp nestled in a valley surrounded by the Coast Range mountains; perhaps the Chinese camp would have contradicted the order that dominates these images (Figure 6.5).[11] Two of the photographs focus on different groups of neat miners' cabins, and these two photographs set up an interesting comparison.[12] The first photograph shows a group of modest miners' cabins across a stream from the larger building in the foreground (Figure 6.6). The cabins are all similar: they have a central door with a window on either side, a small front porch, and a peaked roof. Roughly sixteen by ten feet in plan, the cabins probably had two rooms, no fireplace, and a shared outhouse. About half of the cabins appear to have an addition on the rear, probably

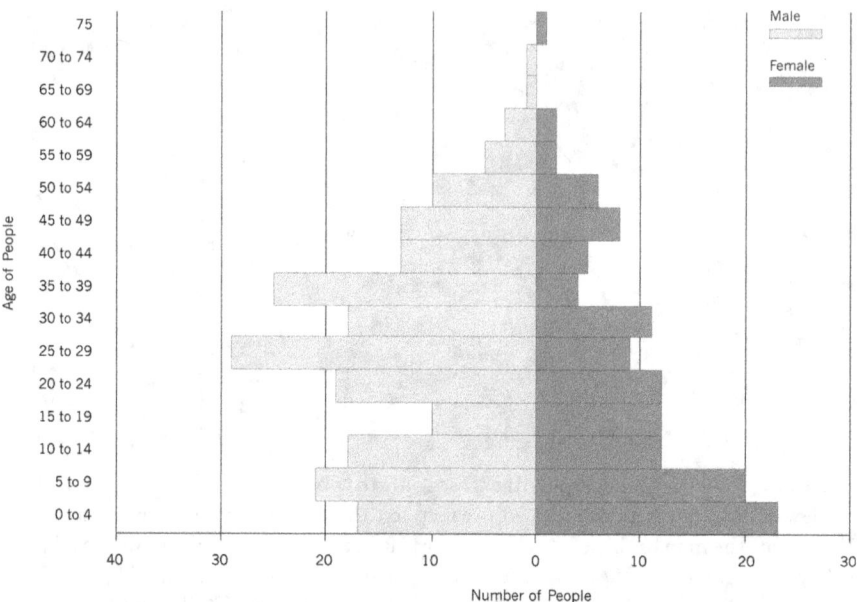

FIGURE 6.3 Population Pyramid for the Guadalupe Mine, 1880, without Chinese Workers
This chart shows the mine community minus the Chinese workers. (Based on the Records of the Bureau of the Census, Manuscript Schedules of Decennial Population Censuses, 1880.)

a kitchen, as evidenced by the visible stove stacks. All the cabins are simple wood-framed structures with board-and-batten walls and shingled roofs, although they are not all identical.[13] All of the cabins face the same way, toward the road, stream, and what is probably the superintendent and mine managers' houses on the opposite hill, approximately where the photographer stood to take the photograph. Some of the cabins have long front stairs, awkwardly built down the hillside, to make this orientation possible. Almost no clutter is visible on the outside of the cabins—only the cabin on the extreme left has anything on its porch.[14] On the superintendent's and managers' side of the river, in the foreground, the roadway is lined by fences, both a milled picket fence and a whitewashed limb fence, and orderly planted trees. In the other photograph, from the east end of the mine, the miners' cottages are in a long row facing the road and then the stream beyond (Figure 6.7). The cabins on the right of this image would look to the reduction works. These cottages were larger than the others, being almost square with a window on the side as well as the two in the front. There is seemingly a one-to-one correspondence between cottage and

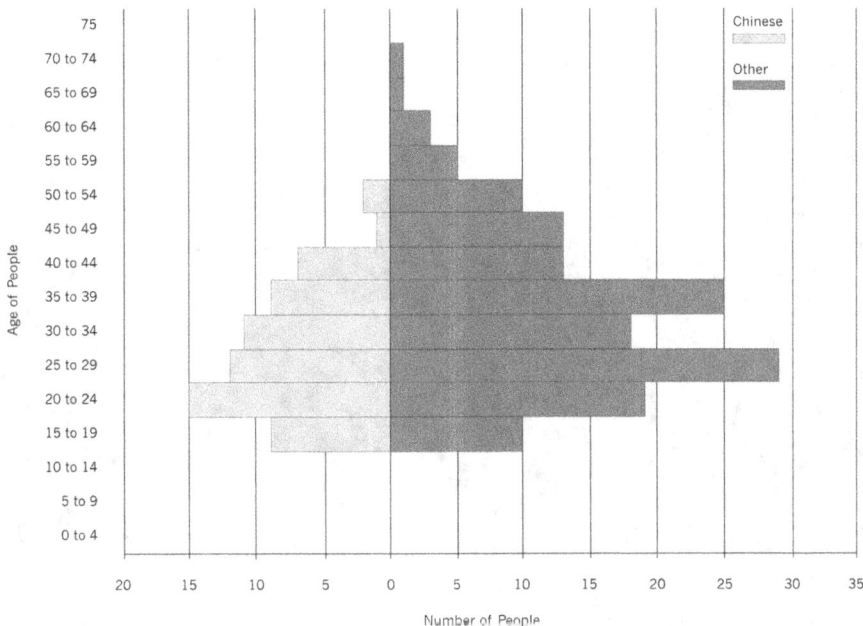

FIGURE 6.4 Worker Population Pyramid for the Guadalupe Mine, 1880
This chart compares the Chinese with the non-Chinese worker population. (Based on the Records of the Bureau of the Census, Manuscript Schedules of Decennial Population Censuses, 1880.)

outhouse. Most prominent, however, are the fences that defined yards and separated most of the cabins from one another, a feature that is not part of the landscape of the cabins in the first photograph.

The physical form of these two groups of cabins points to differences in status and differences in culture. The records do not say who lived in these two groups of cabins; however, the larger cabins of the eastern group in the second photo point to family groups, and being relatively more expensive to build, point to a more privileged population, probably the higher-paid whites. The fences are a telling sign in the landscape: fences of this type defining front as well as rear yards were a common practice in Englishtown at New Almaden, just over the hill from this site. Fences imply control over the space fenced, including psychological ownership, the practical ability to cultivate a kitchen garden, female hominess, and refinement. The western group had much more modest structures closer together and without defined yards. Simple economics point to the less privileged population and possibly a single male population without families, who ate in company dining halls.

RACE, FAMILY, AND CAMP LIFE 221

FIGURE 6.5 View of Guadalupe Village, about 1880
The furnaces and condensing system are in the foreground, and the planilla and shaft house are at the center of the image. The distinct housing areas to the right of the image are readily located on the plan of the Guadalupe Mine (Figure 6.2). The superintendent's house and gardens are to the upper right in this photo. (Courtesy of the Babcock Collection, 1954.016 v.2:45.)

The neat, orderly, and picturesque qualities of the camp point to management's guiding hand. The photographs in the Babcock Collection of the superintendent's house and grounds show that these qualities were well practiced at the top (Figure 6.8).[15] The grounds feature extensive gardens with exotic plant materials, and some photos in the collection even show peacocks.[16] While the superintendent's house itself is difficult to see clearly in the photographs, the bookkeeper's cottage is an expansive structure with an equally extensive house next to it, possibly the superintendent's. Beside these houses to the right of the image is a building with a bell tower that served as both school and church (Figure 6.9). Also, the energy and expense spent on the camp as a whole point to a number of informed conjectures: there may have been a sincere hope and belief that the mine was a long-term investment worthy of substantial improvement. The owners and managers may have made these investments to attract further investment, as well as gaining credibility from mimicking the appearance of the nearby New Almaden hacienda.

FIGURE 6.6 View of Guadalupe Village showing cabins, about 1880
This view shows workers' cabins photographed from the superintendent's grounds across the riverbed. (Courtesy of the Babcock Collection, 1954.016 v.2:46.)

The Redington Mine

The Redington Mine, first developed at the same time as the QMC takeover at New Almaden, did not copy the complex racial hierarchy of the workforce at New Almaden. Instead, the mine was worked by a simple two-tiered workforce composed of Protestant owners and managers and a mostly Irish Catholic workforce.[17] This two-tiered workforce resulted in a markedly different camp landscape than at camps with a more racially and ethnically diverse workforce. The town was located in a broad flat area at the northern end of a long valley in northeastern Napa Company (Figure 6.10). Unlike the separate settlements at New Almaden, Knoxville was one settlement, in keeping with its all-white, all-English-speaking inhabitants. Built right on top of the mine, the town was named for an early superintendent, whereas the Redington name for the mine itself came from the major shareholder of the mine: John Redington. By the late 1870s there were two company stores, offices, a hotel, a Wells Fargo office, a large kitchen, and dormitories located centrally near the mine works. On a hill to the north and east of the mine were the superintendent's house, the church, corrals, and a graveyard. One mine inventory lists a Union Hall.[18] There was no school mentioned at the mine; however, there was a school near the Reed

FIGURE 6.7 View of Guadalupe Village showing cottages, about 1880
These cottages were larger than others at the mine, and each seems to have its own fenced yard and its own outhouse. (Courtesy of the Babcock Collection 1954.016 v.2:43.)

Mine, less than a mile from Knoxville. Radiating from the center were three or four separate areas of cottages for married workers. Built in rows, some of these houses were densely packed, whereas others were spread out. The superintendent's house was furthest from the furnaces; other houses were very close. A rare photograph of Knoxville shows what may be boardinghouses to the left, and to the right a row of cottages, most with fenced yards.[19] In the foreground is a larger, two-story house.

The owners and managers of the mine, local elites in Napa County and businessmen in San Francisco, had close connections to the gold and silver mines of California and Nevada; the superintendents of the Redington Mine often came from the Comstock, and the Redington Mine sold most of its quicksilver to the domestic gold and silver industries. In part, the organization of the Redington Mine was based on the contemporary gold and silver mining industries—at least as far as not hiring Mexican workers, who had been the mainstay of the workforce at all of

FIGURE 6.8 View of Superintendent's House Grounds, Guadalupe Village, about 1880
Mimicking the Casa Grande at the New Almaden Mine not far over the hills, the superintendent's house and grounds at the Guadalupe Mine were elaborate and grand; note the precise plantings. Plant materials are organized on a grid and are spaced in rows according to the grid. Peacocks and honeybees were kept on the grounds as well. (Courtesy of the Babcock Collection, 1954.016 v.2:46.)

the other quicksilver mines up to this time. Instead it hired Irish workers, the lowest tier of "white" worker on the racial hierarchy. Census records show that many of the Irish miners had come to California from Ireland via Australia and Brazil. These miners, part of the migrations from Ireland following the potato famine and its aftermath, may have gone to Brazil and Australia to work in gold mining before eventually landing in California.

Adhering to the racial and ethnic standards of the gold and silver industries, the Redington Mine's owners and managers also adopted the quicksilver industry's reliance on low-cost laborers and a camp that was a company town. Knoxville was a company town—all the land and buildings were owned by the Redington Quicksilver Mining Company. As was reported in the *Mining and Scientific Press* in 1875:

> The company owns the entire town and miles of surrounding lands, with a vast store of all the ordinary and extraordinary necessities; provisions, clothes, medicines, etc.; hotel and stable, shops, and even the church edifice, a very creditable emblem of civilization.

FIGURE 6.9 Church and Schoolhouse, Guadalupe Mine
The Guadalupe church and schoolhouse, shown here some thirty years after the closing of the mine. (Historic American Buildings Survey, Robert Kerrigan photographer, 1936, CA-157-2.)

> Perhaps it is one vast monopoly! Who knows? If so, this fact failed to appear in any form of complaint from their tenants and operatives.[20]

The population pyramid for Knoxville in 1870 shows a heavy bias toward men of working age, from twenty to forty-nine (Figure 6.11). There was only one adult woman for every three adult men, making nuclear families less common than at New Almaden but still a mode of living that was supported by the company. The 1880 population pyramid shows that major changes took place at the mine during the 1870s (Figure 6.12). By 1880 the number of women at the mine had more than doubled, and the number of children had quadrupled. Of the 147 men living at Knoxville in 1880, 45 lived in nuclear households (cottages), 90 lived in the dormitories, and 12 lived as boarders in the cottages.[21] By 1880 there were also about twenty Mexican men in the workforce, about half of whom had wives and children living with them. Company records show that by 1875, at the height of the quicksilver boom, the company had built over fifty cottages and dormitories and a large kitchen to house and feed the workers.[22]

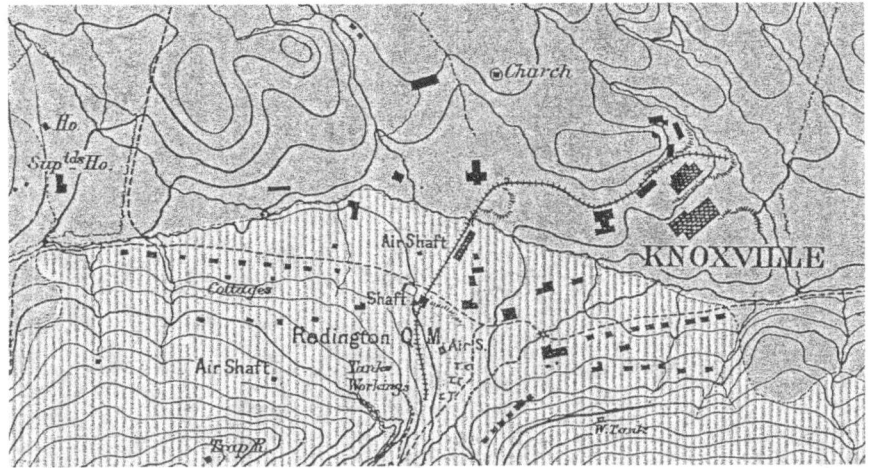

FIGURE 6.10 Plan of the Redington Mine, Knoxville, Napa County 1880s (north is to the left)
The main mine shaft was near the center of the map, with the reduction works to the upper right. Three or four distinct housing areas with cabins are shown. The superintendent's house was on the hill to the left. A graveyard was on a ridge just off the map to the upper right. (California State Mining Bureau map.)

If a Comstock miner made $3.50 a day, minus an estimated $1 a day for room and board, his pay would have been roughly $2.50 a day—compared with $1.80 a day for a single miner at Knoxville.[23] Records for the mine show that in 1875, miners were paid $45 per month and board, whereas laborers were paid $30 per month and board. Roughly 40 to 45 percent of those employed worked underground, although not all men who worked underground were classified as miners. Married miners were paid $20 in lieu of board, and they paid the company $7.50 per month for a cottage. Based on a work month of twenty-five working days, a married miner would have made $2.30 per day, a single miner $1.80, and a single laborer $1.20. Unionized miners in the Comstock Mines made $3.50 to $4 a day at this time, without any allowance for room or board. Perhaps the Redington Mine got by with lower wages by employing Irish miners, who were the lowest-level whites in the racial hierarchy at the time, and the mine company recaptured a good deal of the money that was paid to the workers through cottage rent and purchases at the company store.

Married employees of the Redington Mine may have received some favoritism, both in total compensation, including rent and board allowance, and in the stability of the amount of work they received from the company. The 1880 census had a space for recording how many months a worker had been unemployed in the last year. The year 1880 was a hard one for the mine; however, almost all of the married men living

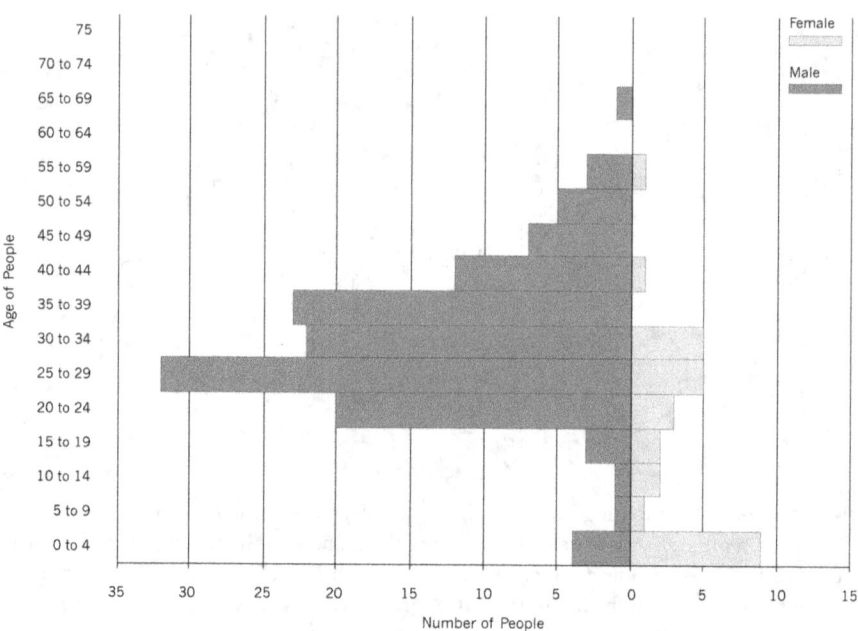

FIGURE 6.11 Population Pyramid for the Redington Mine, 1870

The chart shows a heavy bias towards men of working age, with few nuclear families at the mine. Note the relatively large number of children four and under, perhaps suggesting a new stability for, and acceptance of, families. (This chart is based on Records of the Bureau of the Census, Manuscript Schedules of Decennial Population Censuses, Knoxville, Napa County, 1870.)

in cottages were kept employed during the year prior to June 1880. As an aggregate, these men were unemployed one month in that year. For the men living in the boardinghouse, it was a different picture. Of these ninety men, half were unemployed for an average of 4.8 months in that year. The other half reported no unemployment. The corresponding number at the boardinghouse at the Guadalupe Mine was four months; nineteen of thirty-five men had some unemployment. These numbers speak to an unstable workforce among the boarding class—unemployed men were either staying at the mine hoping for employment or had come to the mine after being unemployed elsewhere. Either way, the married workers at Knoxville and Guadalupe were privileged, with more stability and a greater amount of work received.

Being a monopoly, the Redington Mine kept profits within the company by providing room and board, and it made profits from the company store.[24] With these methods, the Redington Mine company managed to keep its labor costs in line with those of New Almaden and the other old mines. Late in the 1870s, however, as the

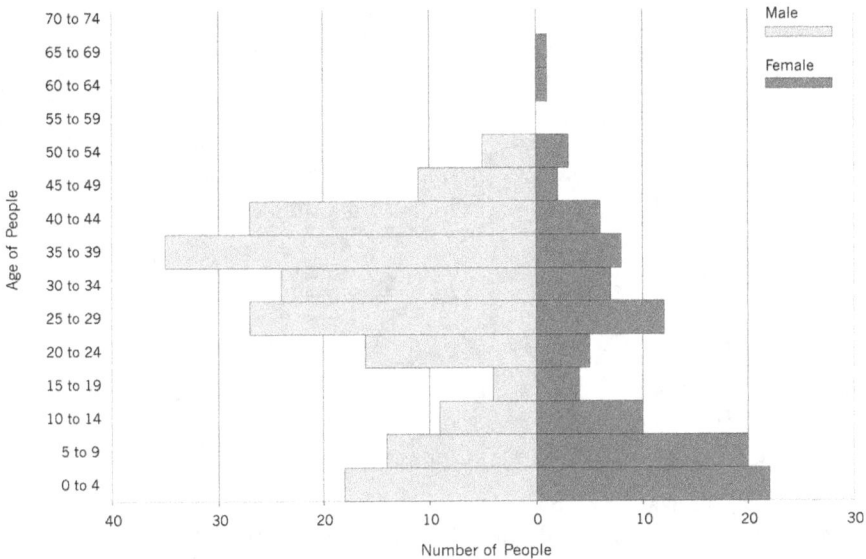

FIGURE 6.12 Population Pyramid for the Redington Mine, 1880

By 1880 the number of women had doubled from the number in 1870, and the number of children had quadrupled. (This chart is based on the Records of the Bureau of the Census, Manuscript Schedules of Decennial Population Censuses, 1880, Knoxville, Napa County.)

mine faltered, Mexican miners were hired as a lower-cost labor solution.[25] The camp organization at the Redington Mine was a hybrid modeled on sources other than the New Almaden Mine. Features and causes of its hybrid state included its simple, two-tiered workforce and its time and place of development, in the 1860s in a very remote area in northern California. Rather than being organized into separate villages, all the workers lived within one settlement, with difference between them visible primarily in whether they lived in cottages or a boardinghouse.

The New Mines

The camps at the new mines were developed based on the need to reproduce their two-tiered workforce of white management and Chinese labor. At least two of these mines—the Great Western and the Napa Consolidated—successfully operated into the twentieth century using Chinese labor and supporting Chinese camps. Although the camps of the new mines were company towns in that they were on company property and the mine company controlled who lived there, they were different from

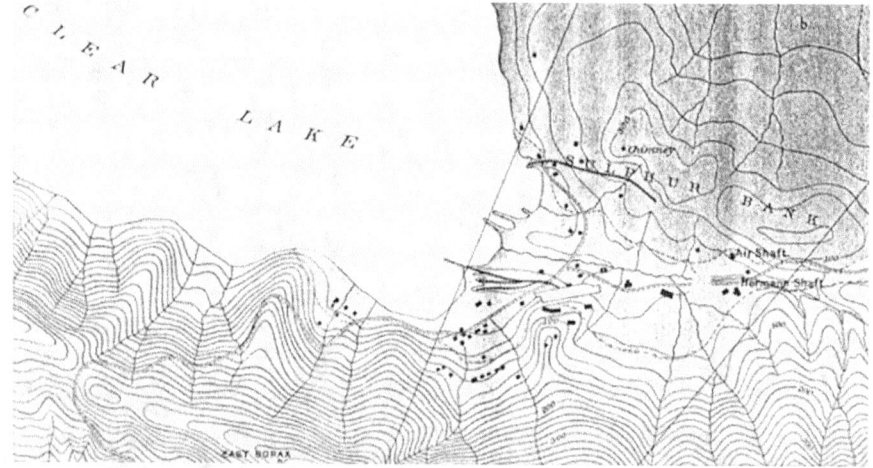

FIGURE 6.13 Sulphur Bank Plan, about 1880

A few houses for managers and non-Chinese workers are shown. The Chinese camps, which would have been extensive, are not shown. At bottom left is a label reading "East Borax." This borax was mined by Chinese miners beginning in the late 1860s. The Indian Rancheria is off the map to the top right. (California State Mining Bureau map.)

the older quicksilver villages and towns in important ways. First, Chinese workers were not charged rent. The common phrase used in reports on the new mines by economic geologists was that the Chinese miners "find themselves." The mine managers would designate an area, often out of the way, where the Chinese miners could build their own settlement. The mine company simply removed itself, theoretically, from any responsibility over housing for the Chinese. In practice, however, issues such as stolen building materials and policing the settlements did burden the mine company. Second, the white population was not isolated at the new quicksilver camps. Most of the new mines were within easy traveling distance of young towns in the Coast Ranges such as Lower Lake, Middletown, and Calistoga. There were still company stores at the new mines, but they were not monopolies for the white populations to the same degree as they had been at the older mines.

The two best documented camps at new mines are in the Clear Lake area (see Figure 7.1): the Sulphur Bank Mine and the Great Western Mine. The Sulphur Bank Mine was the earliest (1872), largest, and most productive of the new mines during the 1870s. The camp spread for a few hundred yards along the bank of Clear Lake (Figure 6.13). The mine works included wharves on the lake for water-borne transport, although most freight (including shipping quicksilver) came to and from Calistoga

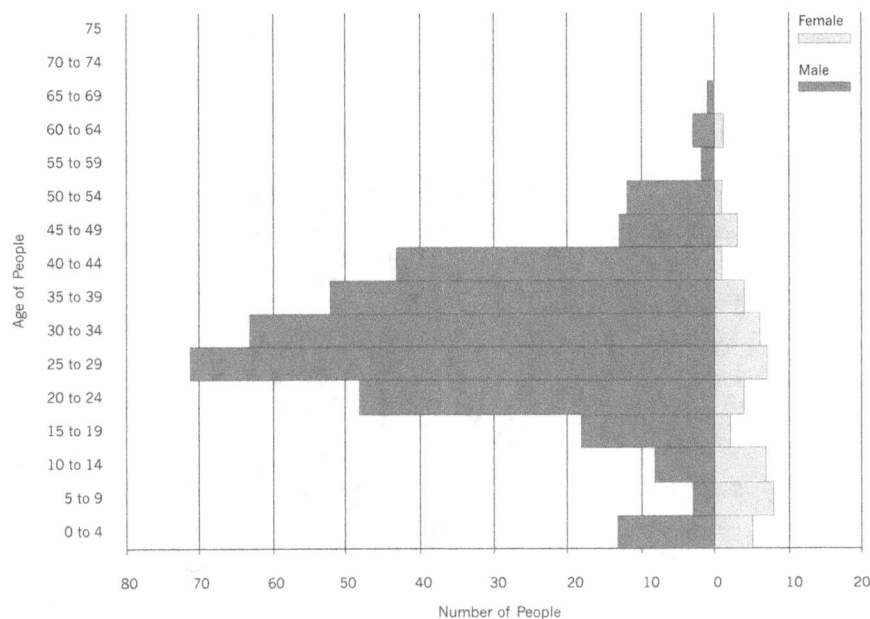

FIGURE 6.14 Population Pyramid for the Sulphur Bank Mine, 1880
Note the large working-age male population, composed mostly of Chinese miners. (This chart is based on the Records of the Bureau of the Census, Manuscript Schedules of Decennial Population Censuses, 1880, Sulphur Bank, Lake County.)

via Middletown. Next to the mine was the long-established Indian Ranchería, although there is no record of Native Americans employed at the mine. Mine housing tended to be on the opposite side of the mine works from the Ranchería in two distinct areas. An 1875 report produced for the directors of the Sulphur Bank Mine lists the following: one hotel with ten bedrooms; two sitting rooms, and two dining rooms; two houses occupied by the families of mechanics; one lodging house for miners with fifteen rooms; one office building with two offices and four bedrooms; and numerous mine buildings.[26] The new mine landscapes were missing the numerous cabins of the families of the miners that were so prominent in the landscapes of the older mines.

The 1880 census listed 325 men working at the mine, of whom 216 were Chinese. The population pyramid for Sulphur Bank shows the sex and age distributions common to the camps of the new mines (Figure 6.14). The working male population is overwhelming due to the 216 male Chinese employees. Twenty-five women and forty-six children were listed in the census, none of whom were Chinese. There were

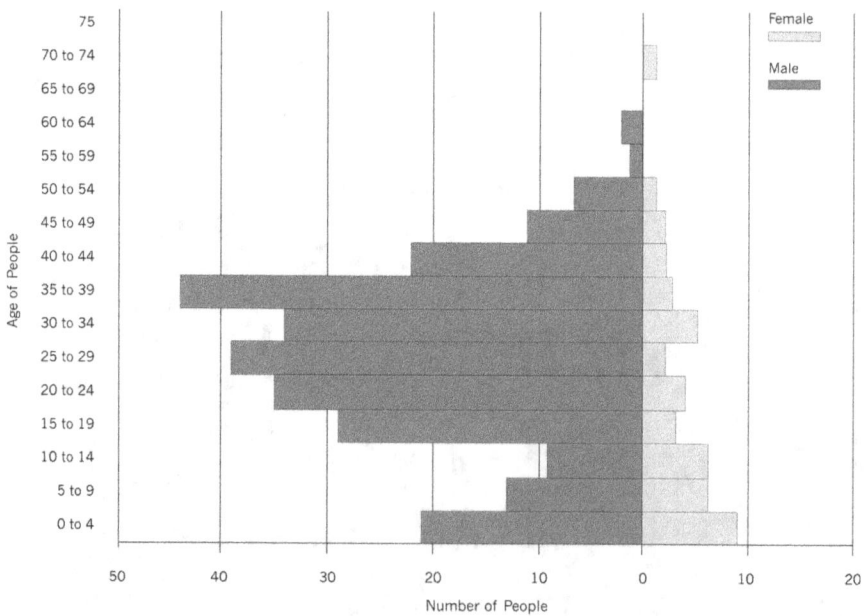

FIGURE 6.15 Population Pyramid for the Great Western Mine, 1880
Note the similarity between the population at Sulphur Bank in 1880 (Figure 6.14) and the population at the Great Western. (This chart is based on the Records of the Bureau of the Census, Manuscript Schedules of Decennial Population Censuses, 1880, Great Western Mine, Lake County.)

twenty nuclear families, none including anyone Chinese. Unlike at the older mines or the other new mines, probable prostitutes were visible in the Sulphur Bank census.[27] Five women, one of whom was forty-nine and the other four in their twenties, lived next to one another in separate houses. Each of the women in their twenties had at least one small child and was listed as married but no husband was present in the census.

At the Sulphur Bank Mine, the Chinese workers "found themselves." The census enumerator listed the Chinese workers in fourteen dwelling units; a nineteen-year-old man named Ah Dorr was listed as the Chinese boss. There were two larger groups of about 35 miners each, nine groups of from 9 to 15 miners each, two small groups of 5 miners, and one pair of miners. One of the medium-sized groups, with twelve men, consisted mostly of nonminers; instead they were cooks or launderers. Housing for the white workers and their families at Sulphur Bank was provided by a contractor: the Getz Brothers Company of Lower Lake. This company ran a store at the mine, had the contract for all the freighting, ran the hotel and boardinghouses, and owned the houses where workers with families lived.[28] This method greatly decreased the amount of capital required to operate the mine. It also meant that the company could not profit as much from the workers, but with mostly Chinese workers the recipients of the profits were limited to the 100 or so white workers and their families, many of whom were managers or in privileged positions and had the resources to move elsewhere if they felt it necessary.

The Great Western Mine was similar in many ways to the Sulphur Bank Mine. However, unlike Sulphur Bank, the Great Western operated steadily from its first production in 1873 through the end of the century under one superintendent, fostering a remarkably stable mining camp. Like Sulphur Bank, the Great Western relied on Chinese labor. The population pyramid for the mine is structured like that of Sulphur Bank, with a massive male working population and a few white women and children (Figure 6.15). The Great Western had a smaller proportion of white to Chinese workers: 21 percent of all workers were white at Great Western, whereas the corresponding figure at Sulphur Bank was 33 percent. Great Western was also more family-oriented among its white working population than was Sulphur Bank; half the white men at Great Western were married, whereas only a quarter were married at Sulphur Bank.

The camp at the Great Western, like that at Sulphur Bank, lacked formalized and expensive company town planning such as at the New Almaden hacienda or the Guadalupe Mine. The long, linear ore body at Great Western resulted in mine buildings that were spread out, rather than concentrated around a single shaft. Living areas for the people associated with the mine were similarly spread out along the ore body. The whites at the camp lived either at the boardinghouse or in cabins throughout the area for families.[29] The boardinghouse had a large room called the "People's Hall" at which functions were held.[30] The Chinese workers lived in one of two Chinese camps located at opposite ends of the Great Western settlement (Figure 6.16). The Napa Consolidated (Oat Hill) map has a clearly labeled Chinese camp (Figure 6.17). At the Great Western the camp located between the superintendent's home and the company store was the No. 1 Camp and was said to have men from the Canton area. Camp No. 2, or the Brown China Camp, was located on the opposite hillside. Helen Rocca Goss, daughter of long-time superintendent Andrew Rocca and author of two books on the mine, related that these men worked separately in the mines, and on at least two occasions there were fights between members of the two groups.[31] Andrew Rocca, the superintendent, described in a letter to his future wife how 125 men were involved in one encounter, "one company against the other."[32] In addition to these differences there were class differences among the Chinese. According to one of Goss's sisters,

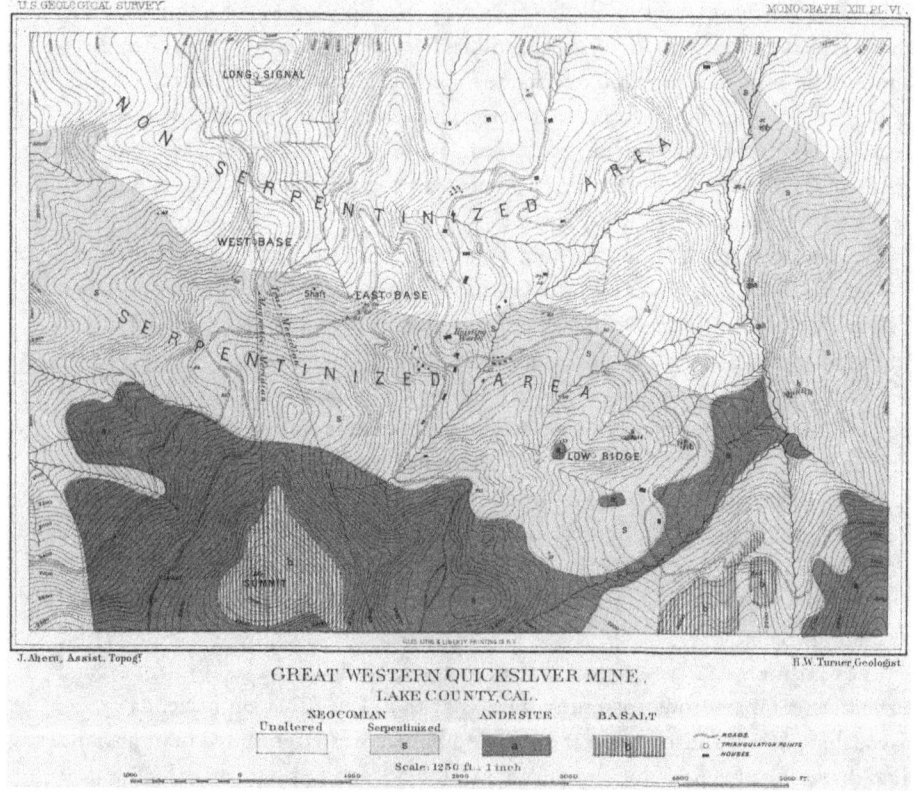

FIGURE 6.16 Plan of the Great Western Quicksilver Mine, 1880s

This plan shows the major elements of the Great Western camp landscape. Chinese camps are not labeled on this map, but based on Goss's description, they are perhaps the grouped small squares in the middle and upper middle of the map. (Becker, *Geology of the Quicksilver Deposits of the Pacific Slope.*)

Those who lived at the Great Western in the eighties and early nineties became well-acquainted with two types of Chinese. There were the Coolies, speaking little if any English, living most primitively in "China Camps," and the more superior, educated men who managed their business affairs, or worked in the home or the store. The miners wore a kind of dungaree costume, similar to the work clothes of sailors, plus the woven straw hats from which I believe their name is derived.[33]

From photographs and descriptions, one sees that the Chinese camps were ramshackle collections of small cabins built in discernable groups suggesting "streets" and "alleys" (Figure 6.18). As Goss wrote of the Chinese camps:

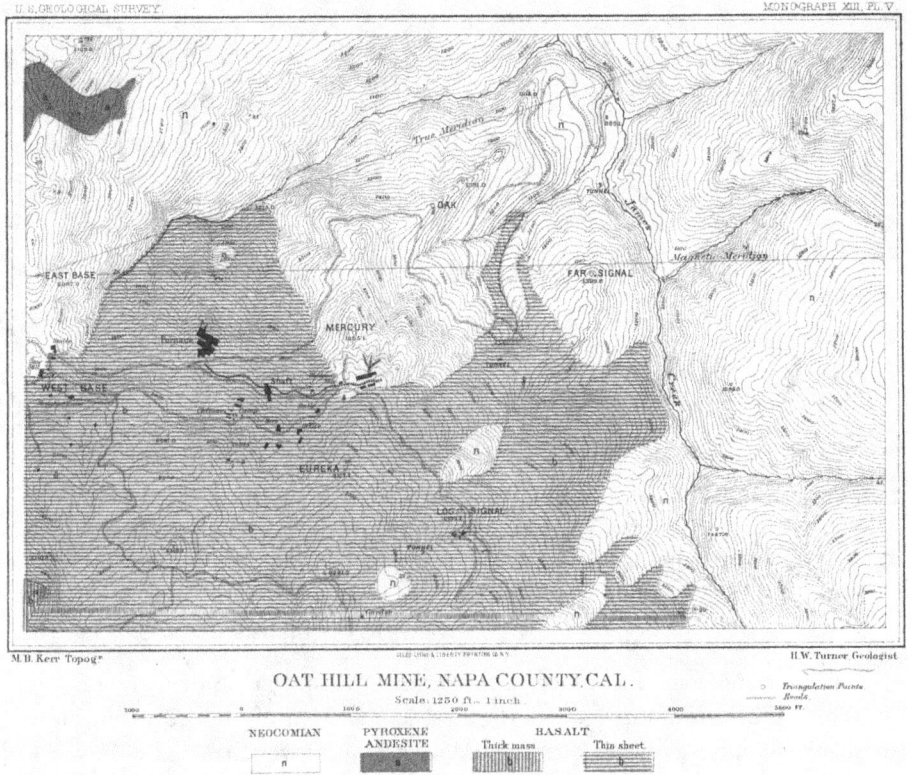

FIGURE 6.17 Plan of the Napa Consolidated (Oat Hill) Quicksilver Mine, 1880s

This plan shows the major elements of the Oat Hill camp landscape. It shows a Chinese camp, center left, with long, thin buildings characteristic of boardinghouses or barracks. The 1880 manuscript census for the mine lists the Chinese miners as living in boardinghouses. (Becker, *Geology of the Quicksilver Deposits of the Pacific Slope*.)

It was a childhood thrill to look into an open door and see a squatting man holding his bowl of rice and using chop sticks; not such a thrill sometimes, to see and have to pass, cross old sows, wallowing in a dirty, muddy ditch . . . At the camp nearest us a few ducks always quacked among the pigs in front.[34]

Goss's sister Florence described the camps as having a few barracks-like buildings of rough timber with "a crazy hodge-podge of shacks built by the men themselves of anything they could pick up . . . scraps of lumber, shingles, broken-up boxes, coal oil tins flattened out, etc."[35] How the Chinese got hold of materials is interesting. A mine the size of the Western would have a great deal of scrap wood. However, much

FIGURE 6.18 One of the Chinese Camps at the Great Western Mine, late nineteenth century
The cabins and shacks shown are small, implying men living in small groups. The structures are also irregular in form and haphazard in overall organization, suggesting little company oversight and the use of scavenged materials. (Goss, *The Life and Death of a Quicksilver Mine*.)

of this wood could be burned in the furnaces. Although clearly there was a decision on the part of management to allow some amount of Chinese procurement of materials, Goss did write that "the ordinary Chinese workman had to be watched constantly to keep them from making off with all kinds of supplies." There must have been a balance of "acceptable" culling of materials that was maintained between mine management and the Chinese, but with what was acceptable heavily controlled by the management. As the *Calistogan* newspaper reported in 1878:

> An examination the other day of the effects of the hundred Chinamen employed at the Great Western Quicksilver mine disclosed the fact that they had nine hundred candles, one hundred and fifty sticks of Giant powder (dynamite), nine boxes of opium, thirty pounds of quicksilver and a quantity of mining tools that they should not possess. All these articles, except the opium and candles had been stolen from the Company; the last two named articles the Chinamen had brought from San Francisco in some manner, which is contrary to the rules and regulations at the mine, which require that all such articles must be purchased in the store at the mine. Superintendent Rocca appears to be much annoyed over this discovery.[36]

A couple of facts about Chinese life at the camps come out in this account. First, the superintendent had the power to search where he wanted in the camps. One can imagine a scene in which white mine employees and possibly local sheriffs searched from shack to shack.[37] Second, in addition to the thefts from the mine, Rocca was concerned that the Chinese workers had brought items into the camp that were not purchased at the company store. The rules and regulations of the mine apparently stated that at least the Chinese workers had to purchase their goods from the company store. Rocca had a particular interest in enforcing these rules and regulations since he owned the company store.[38] The raid on the Chinese camp served to enforce subservience on the part of the Chinese, to curb theft of mine property, and to force Chinese workers to make their purchases at the company store.

The raids of the Chinese camps were not limited to looking for stolen materials. Goss recounts that gamblers and prostitutes often came to the camps on payday. Superintendent Rocca raided the Chinese camps on occasion to stop these activities. Goss's mother wrote in her diary: "Mr. R raided Chinese Gamblers last night. ($115)."[39] The gamblers who came to the mine were described by Goss's sister Lillian as "sleek, dainty-handed, city-looking, foppish, Chinese gamblers."[40] That Rocca felt omnipotent in his raids at the camps is clear from the following by Goss: "Andrew [Rocca] seems to have taken most of his children on such forays at one time or another... If I was ever taken on a 'raiding' expedition I was too young to remember it."[41]

Chinese festivals and celebrations, and rituals such as funerals, were watched with interest by the white community. For the Rocca children, Chinese New Year was an event held on the flat in front of the Joss House. There was at least one Joss house at the mine, located on a hill above the larger Chinese camp. "It was a square, barn-like building with large pictures on the walls of various Chinese rulers and deities, as well as of the devil, in front of which punks were kept burning. The religion of the men seemed to be based more on fear of the devil than on worship of any one god."[42] Trying to make sense of the use of the Joss House led Goss to describe it as both a social hall and a chapel. Goss wrote, "We [the superintendent's children] were taught to consider the Joss House as a kind of lodge hall—something like the Masonic temples or halls about which Papa talked."[43] "We were overwhelmed with presents at that time," observed Goss's sister Florence. "Many people from Middletown, as well as everyone in the camp, came for the fireworks, which for some reason were held in the afternoon when it was all noise and little to be seen. The Chinese passed out gifts of nuts and candy to everyone."[44]

Unlike at the old mines, where Chinese workers did only the lowliest of jobs, at the new mines such as Great Western, the Chinese miners performed all of the work. The level of responsibility carried by Chinese workers—and the high degree of dependence of the managers on the Chinese workmen—created a different environment than at the old mines, where Chinese workers were only the lowest of the low. Chinese miners would have understood this, and if they possessed skills and education they would have sought out the greater (although still modest) rewards available at the new mines. For head Chinese at the new mines, their positions may have provided them with any combination of better pay, working conditions, and living conditions. For Ah Shee, the rewards were $1.50 a day, the "highest paid to any Chinese working underground."[45]

Goss recounted in her books stories of her father's relationships with the headmen of the Chinese camps, with whom close relationships were necessary for operating the mine. As Goss recounted, Ah Shee was the boss of the Brown China camp as well as head timberman responsible for the very important job of supporting the underground workings. Goss related that Superintendent Rocca kept a bank account for Ah Shee and that when he returned to China he had over $6,000 in his account. These sorts of personal encounters, with some measure of decency and respect, were probably repeated many times between the whites and the Chinese, especially elite Chinese, at the mines.

The new mines, by employing highly exploitable Chinese labor, were perhaps more ruthless than the old mines, which operated on paternalist models of mine and camp organization. However, for some Chinese workers, especially the higher class of Chinese worker, the new mines offered opportunities not available elsewhere. With only a two-tiered workforce, the Chinese workers at the new mines were at worst second on the racial hierarchy, not fourth or fifth as they were at the old mines. Also, despite the exploitation, for many Chinese at these new mines the mine communities were safe zones within the larger white society. Chinese moved from Chinatowns in cities and towns to the mining camps, and often were not welcome at points between. The Chinese also had their own "backwoods" routes between nearby Chinese mining camps. If there is an analogy to be made in the social structure and landscapes of the camps at the new mines, it is probably in agriculture; work camps were established to that end in various locations in California in which white managers oversaw gangs of Chinese labor.

The Boom Districts

The two major boom districts were the Pine Flat District in Sonoma County and the Sulphur Creek District straddling Lake and Colusa Counties (see Figure 3.7 for the Pine Flat District and Figure 3.9 for the Sulphur Creek District). The Pine Flat District of Sonoma County was the most frenetic of the boom districts during the quicksilver rush.

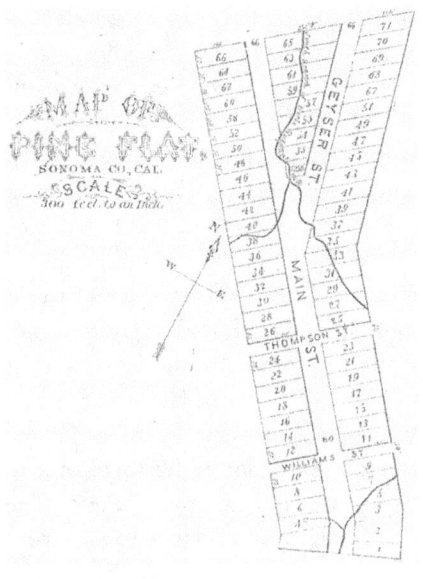

FIGURE 6.19 Map of Pine Flat, Sonoma County

The Thompsons named one cross-street after themselves; Williams Street was named after an eager early investor. (Thos. H. Thompson & Co., *Sonoma County Historical Atlas*, Oakland, CA, 1877, 31.)

In addition to positive press on the value of the cinnabar deposits, what any mining boom district needed were promoters, and the Pine Flat District had identical twin brothers, Granville and Greenville Thompson, who in 1873 staked out a few mining claims in the district, sold at least one at a substantial profit, and claimed a stretch of land that included Pine Flat, the most eligible bit of land for building a town in the area, where "level ground... is like angels' visits."[46] The twins paid for the flat to be surveyed and laid out as a town site. By the fall of 1873, they were building a number of buildings in the town and selling lots (Figure 6.19). The Thompsons also built a store, as well as a hotel and tenement houses for renting to miners; other entrepreneurs built a butcher shop and a number of saloons. By 1874 various mining companies had located their offices in the town, and many miners decided to live in the town and walk to their mine each day rather than live in more primitive conditions at the mine itself. The town boasted thirty-three houses and twenty-four businesses by the fall of 1874, and over sixty houses, eight saloons, and three hotels by the summer of 1875. Interestingly, it seems that Pine Flat also had a bit of tourist trade: one of the livery stables in Healdsburg offered one-day excursions to Pine Flat and the mines.[47] Pine Flat was an easily accessible mining boomtown, heralded as such in local papers, while being close to the population centers of Sonoma County. Tourists visited Pine Flat to experience the much mythologized and fleeting mining boom camp of western lore, in contrast to earlier tourists who visited New Almaden in the 1850s and 1860s to marvel at the scale of industrial development and the richness of the mine.

Newspaper reports referred to the central part of Pine Flat as Americantown. Just outside of town, at opposite ends, were a Mexican settlement and a Chinese settlement.[48] Although mining boomtowns with adjacent minority camps were common, they may have been more pronounced in a quicksilver boomtown because the quicksilver industry as a whole employed a variety of racial and ethnic groups. At Pine Flat the Mexican and Chinese settlements were composed of tents and shanties; Americantown was marginally more substantial with some wood-frame construction.

By early 1875 the boom was almost over, and by 1876 Pine Flat was a ghost town.[49] In 1880 Robert Louis Stevenson and his new wife attempted to settle in Pine Flat: "It was with an eye on one of these deserted places, Pine Flat, on the Geysers road, that we had first come to Calistoga," since "there is something singularly enticing in the idea of going, rent-free, into a ready-made house."[50] Stevenson ended up moving to the nearby ghost town of Silverado when supplies were too hard to come by in Pine Flat. The short-lived quicksilver price boom had brought camps similar to those common in gold and silver mining to the quicksilver industry. Pine Flat, as a boomtown, became a tourist site, a latter-day remembrance of the San Francisco area of the gold rush days. In the course of the industry, however, Pine Flat was pure speculation; investors lost money on the development of the mines, and investors lost money on the development of Pine Flat itself.

The organization of the camp life at the quicksilver mines of California in the nineteenth century constructed and reinforced essentialized racial differences among the laboring groups at the mines. The racialization of camp life was central to the ways people lived, the kinds of buildings they lived and worked in, and the larger landscapes of the mines. There was a significant historical shift in the camps of the quicksilver industry during the 1870s. The shift was from camps being organized by family and community, such as at New Almaden under Barron, Forbes & Co., to being organized to accommodate the labor gang structure of the Chinese workers at the new mines.

The shift in camp organization from family to gang labor was paralleled by a shift in camp typology from full-blown company towns to a situation more like urban factories. The earlier mines provided everything workers and their families needed, and managed to recycle much of the employees' paychecks to the benefit of the company. This system, however, was capital-intensive. In the new mines of the boom era, managers decided to forego the level of potential profits from company stores and land rents of the original mines in favor of allowing workers (mostly white ones) to patronize local towns and otherwise provide for themselves, thus minimizing their investment in the camps. Groups of workers lower in the racial hierarchy, often Chinese, were, however, exploited to the fullest, a policy of mine management that was relatively inexpensive to achieve. In earlier mines, racialization helped to keep workers separate from each other in order to constitute managers' power. White paternalism further controlled the workforce. In the later system, racialization of Chinese workers helped make them dispensable and interchangeable, easy to treat poorly and exploit.

The new mines structured themselves on the new paradigm of Chinese labor. By employing contract Chinese labor, thus reducing the company town aspect of the mines, the new mining companies limited their financial risk. Chinese workers did not have families that needed to be housed, and at most mines the Chinese built their own housing out of scavenged materials. Tax assessments for these new mines of the boom demonstrate the result. The New Almaden and Redington Mines, with little or no Chinese labor and significant company settlements, were assessed at nearly double the rate per flask produced as compared with the younger mines such as the Sulphur Bank Mine or the Great Western Mine, the assessment figures being about seventeen dollars per flask for the older mines, and ten dollars per flask for the younger mines in 1878.[51] The fixed real assets of the older mines, particularly the company settlements, became prohibitively costly when compared with the much leaner new mines. The exception was the original mines of New Almaden and New Idria, where rich and/or plentiful ore made mining profitable even in spite of higher labor costs. New Almaden especially, due to its extraordinary richness, remained an exception in the quicksilver industry and in California, continuing as a closed company town with all the associated abuses. Mary Hallock Foote said of the company stores during her year at the mine in 1876:

> The company stores were run on the ancient system of exploitation—not for the benefit of the company in this case but of the individual owners, and in our time the policy was frank extortion. The prices were exorbitant and the men on the payroll were expected to buy nothing outside which the stores had for sale. A system of espionage kept the manager of the stores informed if anyone infringed this law of the mine... With the Mexicans this rule amounted to peonage—they hadn't the means to move.[52]

Notes

1. The photo is from Goss, *The Life and Death of a Quicksilver Mine*. This is the only photograph of a quicksilver mine community I have seen, and the only photograph I have seen of Chinese quicksilver miners.

2. Just behind the two women standing together in the center of the image are a white man and a Chinese man, almost standing together, but not quite.

3. The 1880 census is the only one that provides a good look at the Great Western Mine population, although the mine was closed by the Parrott case early in 1880 and may not have been back to normal operations by the time the census taker visited in June 1880. The mine was not open for the 1870 census, the 1890 census was destroyed, and by the 1900 census the mine was almost closed.

4. A number of scholars have worked outlining the history of race formation in California. Almaguer, *Racial Fault Lines*. Lisbeth Haas, whose work is more spatial than Almaguer's, has studied identity construction in early California and argues that ethnic and national identities acquired their meanings "through the struggles between contending social groups over who had access to the land and to the rights of citizenship." Haas, *Conquests and Historical Identities in California*, 12.

5. Jessica Sewell has argued that landscapes can be understood by exploring their three aspects: the physical, the imagined, and the experienced. This triumvirate is especially useful in understanding race and ethnicity in the landscapes of the quicksilver camps because of the importance of the imagined for racialized space. Jessica Sewell, *Women and the Everyday City*, xiv–xxii.

6. The nativity count from the 1880 manuscript census for workers at the mine is as follows: Americans, 21; CES, 67; Irish, 4; other European, 16; Californian with Spanish surname, 16; Mexican, 24; Chilean, 4; Chinese, 78.

7. The Guadalupe Mine Collection, 1854–1921, offers no descriptions of the Chinese camp at the mine. The Chinese miners did have deductions made from their pay for purchases at a company store. However, this deduction was made as one lump sum for all Chinese workers, reinforcing the fact that the Chinese workers were not treated as individuals, but as a group.

8. Records of the Bureau of the Census, Manuscript Schedules of Decennial Population Censuses, 1880, Almaden Township, Santa Clara County, Enumeration District 255, William S. Taylor, enumerator.

9. All Chinese at the mine were working-age males from the ages of fifteen to fifty-four.

10. These photographs may have been taken by California photographer Winn Bulmore.

11. By 1884 the Guadalupe Mine was closed, and for the year or two before, the mine was gradually shedding workers. These photographs may have been taken when the camp was only partially inhabited.

12. The manuscript census enumerators most often walked from house to neighboring house collecting information. At Guadalupe the race and ethnicity of households next to one another in the census tend to run in consistent groups. Although this analysis is by no means positive proof of these patterns, it is possible that workers tended to live in groups with other workers of similar status.

13. Account books from the mine show that the cost of the materials for a cabin was seventy-five dollars.

14. Again, the neatness of the cabins could be because the residents had moved out by the time of the photograph. It is also possible that the superintendent demanded neatness.

15. Babcock Collection, Bancroft Library, University of California, Berkeley.

16. The Guadalupe Mine papers include receipts for exotic plant materials.

17. The 1870 manuscript census shows that some of the Irish at Knoxville had lived earlier in Brazil and Australia (based on children listed as born in these countries). The Irish were available in the 1860s and 1870s as a labor force that was both "white" and undesirable, as we saw at New Almaden.

18. Irving Murray Scott Correspondence, 1871, manuscript, OCLC no. 503173076.

19. This photograph is credited to the Sharpsteen Museum in Calistoga, California.

20. *Mining and Scientific Press*, January 16, 1875.

21. Records of the Bureau of the Census, Manuscript Schedules of Decennial Population Censuses, 1880, Knoxville, Lake County.

22. Horatio Livermore, *Annual Report of the Redington Mine*.

23. Randol, *Papers Relating to Quicksilver Mining, Ca. 1849–1894*, vol. 2, folder 2, "Report for Annual Meeting of the Stockholders, 1876."

24. The cottages for married men were listed in the Annual Report as costing $300 each. Rent collected for each cottage totaled $90 per year, meaning that the initial construction cost of the cottages was paid off in under four years. Livermore, *Annual Report of the Redington Mine*.

25. No data exist in the records for comparing wages between Irish and Mexican miners.

26. Ashburner, "Report on Sulphur Bank."

27. The New Idria census for 1860 also shows probable prostitutes living in one of the mine camps. Eight women, all listed as washerwomen, are listed in separate residences side by side. Six had at least one small child, and all were listed as unmarried. When Brewer visited New Idria, he also mentioned prostitution. On the drive to the San Carlos Mine, Brewer saw a house perched on a cliff above the road. On inquiring what it was, he was told, "Oh, a billiard saloon and drinking house . . . A man recently built it and I believe he has other refreshments for the miners, a load of *squaws* went up a day or two ago." In the 1860 New Idria census, there are large Indian groups living near the mine. Brewer, *Up and Down California in 1860–1864*, 139.

28. The *Lower Lake Bulletin* of July 19, 1879, reported that the house of J. E. Tucker, assistant superintendent, burned down, at a loss of $1,500 to Getz Brothers. An article in the *Bulletin* on January 2, 1878, states that Getz Brothers ran the store at the Sulphur Bank.

29. No church was listed at the mine. However, in 1895 Mary Rocca and a number of new converts from a series of revival meetings at the mine founded the Cinabria Baptist Church. Goss, *The Life and Death of a Quicksilver Mine*, 52–53.

30. In 1878 the *Independent Calistogan*, July 24, 1878, reported on a number of major events at the hall, including a Christmas party and balls, one of which was a fundraiser for the mine's "Base Ball Club"; *Independent Calistogan*. The other events were described in ibid. on April 24, 1878 (a picnic), and December 25, 1878 (Christmas party).

31. Goss, *The Life and Death of a Quicksilver Mine*; Goss, *Gold and Cinnabar*.

32. Rocca said that four or five of the men were badly cut, and he succeeded in ending the encounter only by brandishing a rifle. Goss, *The Life and Death of a Quicksilver Mine*, 70.

33. Goss, *Gold and Cinnabar*, 233.

34. Ibid., 233.

35. Ibid., 109.

36. *Independent Calistogan*, September 11, 1878. Goss discusses this article in Goss, *The Life and Death of a Quicksilver Mine*, 81. She states that the article was written in a confused way and that opium was not sold at the company store.

37. Goss wrote that the *Independent Calistogan* of November 5, 1879, carried a brief stating that forty pounds of quicksilver were found on the train in a Chinaman's luggage. Goss, *Gold and Cinnabar*, 520.

38. It was common for mine superintendents such as Rocca to be part owners of the company store.

39. Goss, *Gold and Cinnabar*, 241.

40. Ibid., 241.

41. Of his memories of such an event, Goss's brother Andrew wrote: "I went through the No. 1 camp near our home with Dad on several occasions, and once they were gambling . . . Dad took his cane, scattered money and beans all over the place and clubbed the offenders over the head." Ibid., 241.

42. Goss, *The Life and Death of a Quicksilver Mine*, 89. This was written by Goss's sister Florence.

43. Goss, *Gold and Cinnabar*, 243.

44. Ibid., 244.

45. Goss, *The Life and Death of a Quicksilver Mine*, 69.

46. *Russian River Flag*, July 9, 1874, as cited in ibid., 58.

47. Ibid., 66.

48. Pelanconi, "Quicksilver Rush of Sonoma County, 1873–1875," 70.

49. Many of the mines were involved in litigation; early claimants from the 1860s rush took the opportunity of the new law extending the time to prove one's claim to challenge the rights of the more recent claimants. This litigation was big news through the first half of 1875, and surely it scared away capital investment in the mines. But most important to the demise of the boom was the drastically falling price of quicksilver. Development money for the mines ran out by the end of the summer of 1875, and the mines were mostly abandoned.

50. Robert Louis Stevenson, *The Silverado Squatters*, as cited in Pelanconi, "Quicksilver Rush of Sonoma County, 1873–1875," 98.

51. "California Quicksilver Mines," *Independent Calistogan*, August 14, 1878.

52. Foote, *A Victorian Gentlewoman in the Far West*, 126.

In the early 1870s, with the breakdown of the quicksilver combinations, Thomas Bell, the inheritor of the Barron, Forbes & Co. mercury empire, saw his control of the industry disappearing. Although he still made significant wealth related to mercury during the quicksilver boom and bust of the 1870s, neither he nor anyone else ever controlled the industry to the same extent as before the boom; never again was mercury wealth to be so concentrated in so few hands and in a small number of mines. The racial and ethnic restructuring of the industry during the quicksilver boom was the end of the colonial mode of control that Barron, Forbes, and their descendents had profited from so handsomely. All four models of mercury mining that replaced the colonial model followed one racial hierarchy. The new mines were able to operate with much less infrastructure development than the original mines because they did not need to provide as much support for their Chinese workers, due to their low position on the racial hierarchy. By the early 1890s, the primary market for quicksilver was removed with the introduction of the cyanide processing of gold and silver. In spite of the decline of the industry, mercury mining had lasting effects on the racial and physical landscapes of California.

Conclusion

The Legacy of the Quicksilver Landscapes of California

The Decline of the Mercury Industry in California

By the end of the 1870s the mercury industry in California was in decline. Two court cases in California in the early 1880s—*In re Tiburcio Parrott*[1] and *Edwards Woodruff v. North Bloomfield Mining and Gravel Company*—greatly impacted the industry and hastened its fall. In 1880 *In re Tiburcio Parrott* placed the quicksilver industry at the forefront of the debate around Chinese labor and immigration, affecting the labor pool for mercury mining; in 1884 *Edwards Woodruff v. North Bloomfield Mining and Gravel Company* removed much of the local market for mercury by greatly limiting hydraulic mining.

The Parrott case has an interesting history. At five o'clock on the evening of Friday, February 20, 1880, Tiburcio Parrott, the illegitimate son of John Parrott—one of the richest men in San Francisco—was arrested for violating a provision of the new California Constitution. Being a member of a powerful family in the city and the state, and having lived his life exercising his family privilege, Parrott probably felt he didn't have much to fear in challenging the new "Workingman's" constitution. Nor was Parrott overly frightened when a few days later Denis Kearney, president of the Workingman's Party, tried to raise enough money by subscription to build a gallows to hang him on the "sand-lot" for his direct challenge to the constitution. These events made headlines in newspapers in San Francisco and throughout California. The *Daily Alta*'s article against the proposed hanging enumerates the contributions John Parrott had made to the development of California, and argues that in light of these contributions hanging his son would be "un-American."

The legal battle following Parrott's arrest, which in its resolution was of great importance in the state and to the nation, involved Article XIX of the new constitution, passed by California's voters in May 1879. Article XIX read in part: "No corporation now existing, or hereafter formed under the laws of this state, shall, after the adoption of this constitution, employ directly or indirectly, in any capacity, any Chinese or Mongolian. The legislature shall pass such laws as shall be necessary to enforce this provision." In order to enact Article XIX the legislature amended the penal code by adding two new sections that were approved on February 13, 1880. Section 178 charged any employee of a corporation that employed any "Chinese or Mongolian" with a misdemeanor punishable by fine, imprisonment, or both. Section 179 charged any corporation employing any "Chinese or Mongolian" with a misdemeanor, punishable by fine, and on second conviction by forfeiture of its "charter and franchise, and all its corporate rights and privileges." Tiburcio Parrott, it was charged in the warrant for his arrest, as president of the Sulphur Bank Quicksilver Mining Company, "did willfully and knowingly employ as a workman upon the works of the corporation a certain Chinaman and Mongolian, one Wong Ah Sing, who is a native of China and a subject of the Emperor of China."[2] Nobody, including Parrott, argued against these facts. The case, heard a few weeks later in San Francisco by the United States Circuit Court of Appeals, involved a direct challenge to the constitutionality of the anti-Chinese provisions of the new constitution, a challenge based in ideas of class and the exercise of the power of capital. The new constitution, some contended, born of the Workingman's Party's rise to power beginning in 1877, benefited labor against capital.

It can be argued that it didn't really matter that the test case happened to involve a quicksilver mine, as Chinese workers were employed in many industries in California at the time. The legal battle concerned only the constitutionality of a California corporation's employment of Chinese workers, no matter if they were miners, cigar makers, or weavers. It can also be argued that it really didn't matter that the man under arrest was Tiburcio Parrott. However, I argue that it is significant that a quicksilver mine was the test case. By 1880 the quicksilver mining industry had become dependent on Chinese labor, and the Sulphur Bank Mine, in the early 1870s, was one of the first corporate mines to hire hundreds of Chinese workers to do work that white laborers would not. Equally significant is the fact that the defendants were Sulphur Bank Mine and Tiburcio Parrott, one of the more dynamic combinations of personalities and quicksilver mines in the history of the industry.

Tiburcio Parrot made himself a presence in the quicksilver industry during the quicksilver boom. His father was fortunate enough to have been one of the original group of owners (from the late 1840s) of the New Almaden Mine. Involvement with California quicksilver had helped the family solidify its fortune, and for whatever personal or financial reasons Tiburcio Parrott headed a new round of Parrott family interests in quicksilver. In October 1875 five investors—including John Parrott; two business associates, William and James Burling; H. S. Lightner, who served as the first mine superintendent; and Tiburcio Parrott, as the president—formed the Sulphur Bank Quicksilver Mining Company and bought control of the mine. In 1874, Tiburcio Parrott had also leased the Great Eastern Quicksilver Mine in Sonoma County, at the time a promising prospect.

The Sulphur Bank Mine, together with the Napa Consolidated (Oat Hill) Mine and the Great Western Mine (both about twenty miles south of Clear Lake), had an economic reliance on Chinese labor, which could be had for about half the cost of white labor, if white labor was available in rural areas (Figure 7.1). California's quicksilver mines were the major opportunity for Chinese miners to labor for corporate mines and use skills developed in building the railroads or in other mining work. Comstock miners had successfully organized against the Chinese, but in the remote and modest towns and camps of the Coast Ranges, hundreds of Chinese labored in quicksilver mines, where there was no organized protest by white labor. Through the late 1870s, the Sulphur Bank Mine and the Great Western each were worked by 200 to 250 Chinese laborers, with twenty to thirty white managers and supervisors. The Oat Hill Mine, in the 1870s, worked about half the numbers of the other mines, though later in the nineteenth century the mine employed many more laborers and became the largest producer of the three (Figure 7.2). By using Chinese labor and forcing these workers to house themselves, the mining companies limited their financial risk. Additionally, these mines had harsher work conditions than the Knoxville-area mines, and less investment by the companies to alleviate these conditions reduced their costs.

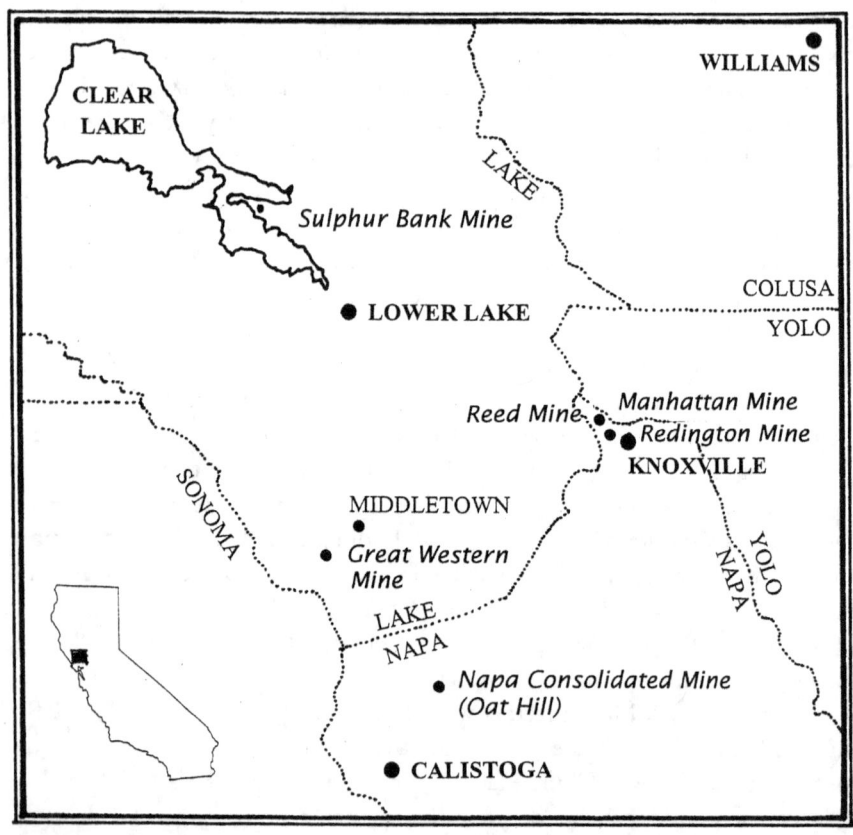

FIGURE 7.1 Map of Northern Mines
Map of the quicksilver mines of the Clear Lake area. In 1872 the Napa County / Lake County border was redrawn under pressure by Napa County to bring the Knoxville-area mines back into the county. (Map drawn by author.)

The Sulphur Bank Mine lay at the edge of the southeast end of Clear Lake, in Lake County (see Figure 6.13). Chinese miners had first worked the site for sulphur, then borax, in the early 1870s. During the quicksilver boom the Sulphur Bank was worked for quicksilver, which proved to be much more valuable, though sulphur continued to be produced. The quicksilver ore at the Sulphur Bank was one great ore mass roughly 1,800 by 900 feet, and about 50 feet thick (Figures 7.3 and 7.4).[3] The cinnabar ore was well mixed with sulphur-rich materials, which forced the mine superintendents to do a great deal of experimenting with sorting ore and concentrating cinnabar. Processing the ore at the site posed problems not encountered at other quicksilver mines and

FIGURE 7.2 Mock Bing Yer

The photograph is of Mock Bing Yer, a Chinese man whose report for the immigration office states he lived at Oat Hill from 1879 to 1894. He is listed in the records as part of Yer Hop & Co., with $1,000 in assets. (From the National Archives and Records Administration, San Bruno, California.)

involved a great deal of experimentation, though visiting geologists and metallurgists commented on a lack of coordinated experimentation. Throughout its operation, from the mid-1870s to its close in 1883 (the end of the Parrott era), the mine had a vast array of nearly every quicksilver reducing technology available.[4] By the late 1870s the mine also had hired a separate company to set up a facility to concentrate the cinnabar ore. This company worked thirty men, mostly Chinese; it took the ore as mined, processed it to achieve a concentrate, and then delivered it back to the mine for reduction in the furnaces. Other services were also contracted out. The Getz Bros. Company, which ran stores in the town of Lower Lake, had the contract for freighting supplies and products between the mine and Calistoga. Getz Bros. also ran the boardinghouse at the mine and the boardinghouse dining room, and owned houses that were rented to the white managers at the mine. Through these two arrangements, the owners of the Sulphur Bank Mine limited their financial exposure.

When the voters of the state of California ratified the new constitution in May 1879, the vote in the town of Calistoga was 126 for, 80 against. At the Sulphur Bank Mine the vote was 2 for, 28 against. Although the ratification of the constitution was a warning shot, the actual act that amended the penal code to prohibit the employment of Chinese by California corporations was passed and signed on February 13, 1880. On February 14, the managers of the Great Western Mine dismissed their Chinese laborers. The superintendent, Andrew Rocca, wrote to his future wife, Mary: "I had to discharge all the Chinamen today. Suspend all operation and do not know

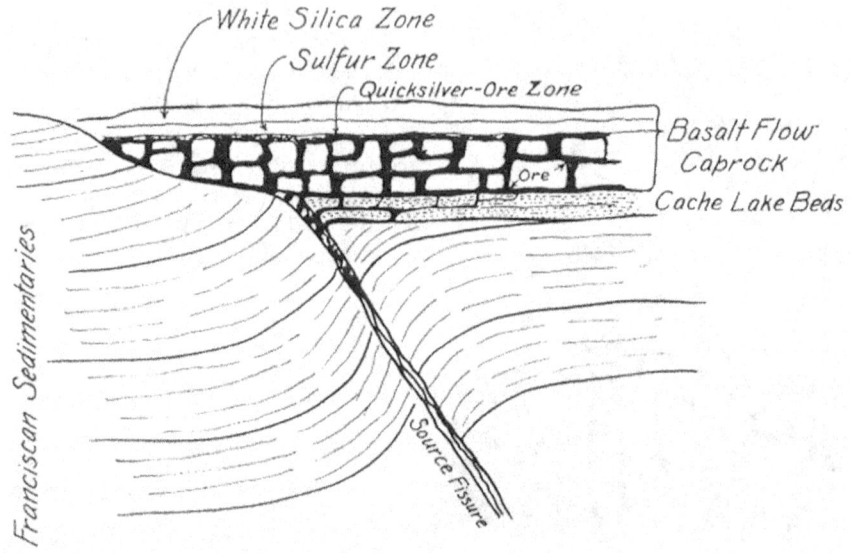

FIGURE 7.3 Section through a Surface Ore Deposit at Sulphur Bank

Sketch of the components of the upper layers of the Sulphur Bank ore body. Volcanic basalts capped sedimentary layers, mostly sandstone, which were impregnated in chambers by cinnabar. (Schuette, "The Geology of Quicksilver Ore Deposits.")

what labor I will get when I start up again, but think I may have to go South and get some negroes to take the Chinamen's place."[5] The Saint John Mine near Vallejo discharged its Chinese workers at about the same time. However, as of February 25, 1880, reported the *Independent Calistogan*, the Oat Hill Mine had not discharged its Chinese workers.

One week after the penal code amendment, Tiburcio Parrott was arrested. The warrant was issued after a man named William H. Barton filed a complaint against Parrott in the court. As soon as Parrott was under arrest, he, through his counsel, applied to the Circuit Court for a writ of habeas corpus, which was granted, and he was released on $500 bail. On February 28, 1880, Superintendent Fiedler of the Sulphur Bank Mine told the *Lower Lake Bulletin* that he would discharge Chinese workers as white workers were hired. Interestingly, Getz Bros. ended its contracts with the Sulphur Bank Mine at exactly this time, and the mine company under Fiedler took over the mine store.

The court case began Saturday, March 8, 1880. The lawyers for Tiburcio Parrott argued that the provisions of the new California State Constitution were in violation of the Fourteenth Amendment of the Constitution of the United States and the law

FIGURE 7.4 The Western Cut, Sulphur Bank Mine
The "Western Cut" at the Sulphur Bank Mine, with hot springs laden with sulphur. At "the Western Cut there are a number of warm springs with considerable excess gas escaping with the water—the whole having the appearance of a series of boiling caldrons." (Bradley, *Quicksilver Resources of California Bulletin No. 78*, 66.)

passed to enforce it known as the civil rights law. Additionally, the new constitution was in violation of the Burlingame Treaty between the United States and the Chinese Empire. The state argued that the legislature was free to amend the laws that permitted corporations within the state. The decision by Judges Sawyer and Hoffman was that the provisions to the constitution in question violated the constitutional rights of the shareholders of the corporations to use (gain value from) their holdings. As to the Burlingame Treaty, the judges stated that although Chinese migration "was a menace to our peace and our civilization," the treaty existed, and the constitutional provisions in question took from the Chinese their livelihood, and thus their right to life. Both judges agreed that the United States treaty was superior to the laws of any state and that the remedy for Chinese immigration was not with the state, but with the general government.

Tiburcio Parrott had won his case. His powerful social position both allowed him to hire the top legal help available and made him largely untouchable. The Parrott family had a long history of lucrative connections with China and the Chinese. The Parrott Building, completed in 1852, was constructed of Chinese granite shipped to San Francisco and assembled by Chinese workmen. Parrott and Co., the shipping agent and banking house founded by John Parrott, was closely tied to firms in Canton and Hong Kong. Tiburcio Parrott and the Parrott family made money from shipping goods, including quicksilver, to China. The Parrotts also made money from importing Chinese goods, and perhaps Chinese immigrants. In California the family made money off the labor of Chinese workers, and made money from selling to them the Chinese goods Parrott and Co. had imported. Given the strength of Tiburcio Parrott's financial ties to China, it was not surprising that he was willing to challenge the anti-Chinese clauses of the California Constitution.

Parrott's ties to China do not entirely explain why it was the quicksilver industry in particular that defied the constitution. The quicksilver industry was central to the question of Chinese labor for several reasons. First, it was a rural industry, located in the Coast Ranges far from urban centers. It was also, by the early 1880s, a modest and unstable industry that did not attract the attention that larger, more stable industries did. The remote mines did not incite the urban crowds, particularly the "sand-lot" crowd. The quicksilver industry itself had structurally changed during the quicksilver boom to a Chinese-dependent industry, a fact that created much lower costs of production for the mine corporations. The Sulphur Bank Mine was a unique combination of the nastiest mine of all the quicksilver mines in the state, which made it unable to get white workers, and a profitable mine that could be realized only through Chinese labor.

The whole affair of the case contributed to the general shift in power away from the Workingman's Party toward the interests of capital occurring in the early 1880s. After the trial the mines continued much as they had before the challenge. A few months after the trial, the *Independent Calistogan* rather resignedly stated, "thirty moon-eyed lepers came to town last Friday on their way to the Napa Consolidated Quicksilver Mine ... The Chinese seem to be almost indispensable in a quicksilver mine, though we hope they never find their way into a mine in this vicinity."[6] Although the closure of many mines during the trial caused a slight rise in the quicksilver markets, the early 1880s proved to be a time of depression for the industry. The Parrotts stopped operations at the Sulphur Bank Mine in 1883, although both the Great Western and Napa Consolidated Mines continuing operating for many years afterward.

Whereas *In re Tiburcio Parrott* ruffled the quicksilver industry, *Edwards Woodruff v. North Bloomfield Mining and Gravel Company* posed a deeper challenge. This case revolved around the environmental effects of hydraulic mining, specifically the vast amounts of debris that were washed downriver from the mines and that changed and blocked waterways, resulting in flooding. Farmers in the Sacramento Valley were particularly impacted, and one farm owner, Edwards Woodruff, filed suit in September 1882 claiming that hydraulic mining was a public and private nuisance. For Thomas Bell this case threatened his sales; hydraulic mining used a great deal of mercury. More important for Bell, though, was the potential loss of influence and power in the western gold mining industry that his control of mercury supplies had enabled him to gain.

Thomas Bell and the Board of Council of the Miners' Association of California published a pamphlet with a statement counterarguing the benefits of mining for both the past and future development of the state.[7] They also argued that the major hydraulic mines employed fewer Chinese laborers by percentage than many other industries in the state, including agriculture. They were arguing, they said, against claims that the mines were owned mostly by foreigners, and that they were run only with "Chinamen, powder and water."[8] The association, agreeing that farmers had been injured by mining activities, proposed engineering solutions to the "debris question," solutions they argued would aid farmers while not hurting miners or mining towns.

Judge Lorenzo Sawyer, in 1884, delivered his opinion in what has become known as the "Sawyer Decision," a decision that effectively ended hydraulic mining. Following two years of litigation, this decision prohibited the discharge of mining debris in the Sierra Nevada. For Thomas Bell and his partners in the North Bloomfield hydraulic mine, this case was devastating. Bell—besides being a major owner of hydraulic mines, including the North Bloomfield—had spent money on the case and openly lent his name to arguments against it. The decision effectively eliminated one of Thomas Bell's carefully cultivated uses for mercury as part of his integrated system of control of the industry.

The last year of high production figures for California's quicksilver industry was 1883. From a high of 79,395 flasks in 1877, production decreased to 46,725 by 1883, then fell precipitously to 31,913 in 1884. During 1883–84 the Redington, Sulphur Bank, and Guadalupe Mines all shut down. The causes of the market decline included decreased Californian demand and a glut of quicksilver on world markets, coupled with price competition from the European mines.[9] Thomas Bell curtailed production at New Idria as he again sought combinations to control the quicksilver markets and did not want large quantities of mercury from New Idria to try to sell. The boom districts had also all but disappeared by the early 1880s. The only mines left operating after 1883 were the New Almaden Mine, still with plenty of good ore, and the two most successful new mines, Great Western and Oat Hill. New Almaden dismissed

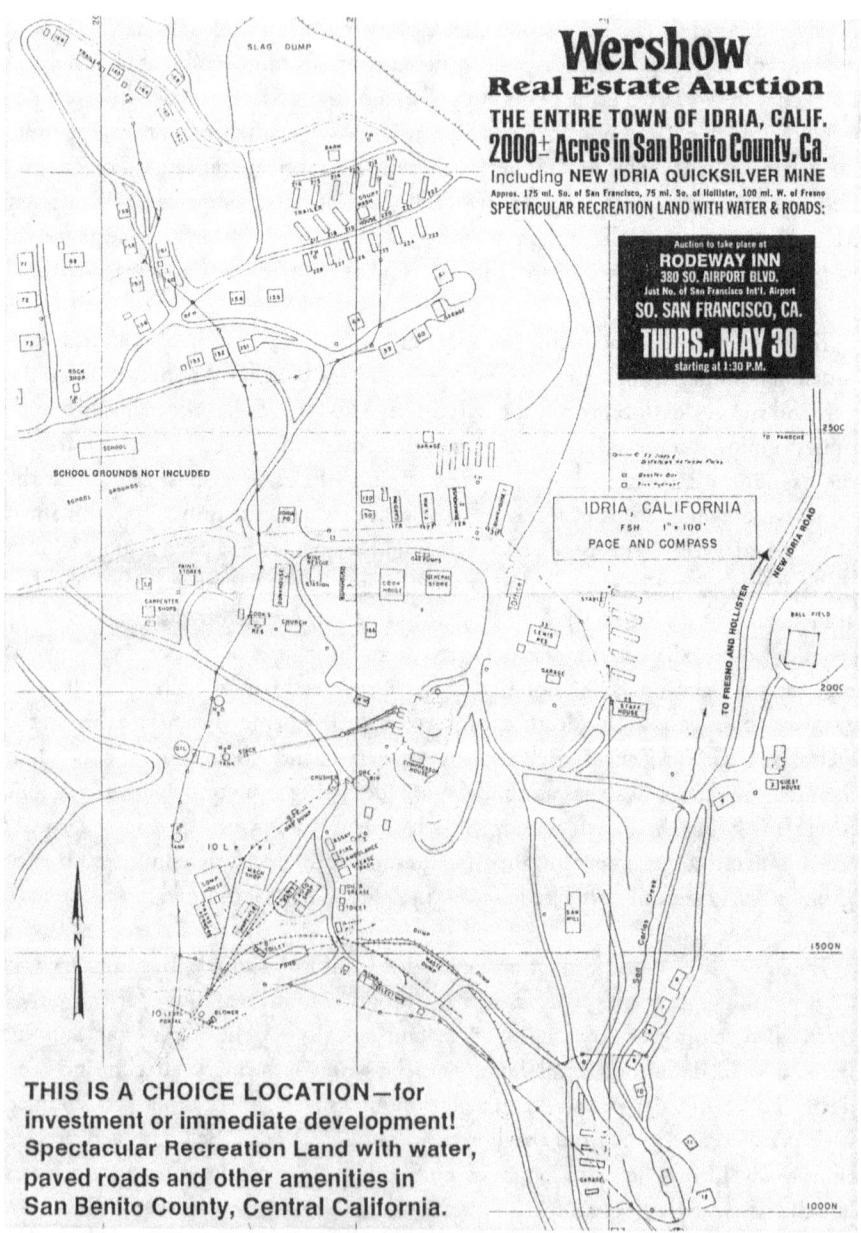

FIGURE 7.5 Flyer for the New Idria Real Estate Auction
This plan shows the town as it was sold at auction in 1973. The main entrance to the mine is at bottom left. The center of town is at the center of this image. (Author's collection.)

its Chinese laborers in 1883, but Great Western and Oat Hill retained their Chinese workers, and the communities that these miners built at these mines continued as Chinese outposts in the California countryside into the twentieth century. In the early twentieth century, even these California mines were exhausted (except New Idria, which has never been fully exploited and retains supplies of ore). When quicksilver mining in the United States moved into the Big Bend of Texas in the early years of the twentieth century, it was Mexicans displaced by the Mexican Revolution who became the mainstay of the workforce.[10]

The Quicksilver Legacy

On May 30, 1973, a unique real estate auction was held near San Francisco. For sale was "The Entire Town of Idria, Calif. 2000 (plus or minus) Acres in San Benito County, Ca. including New Idria Quicksilver Mine" (Figure 7.5).[11] First prospected in the 1840s, commercially developed in the 1850s, and controlled and manipulated by William Barron, Thomas Bell and the Bank Crowd, the longest-operating mine in the history of California and the second-largest producer of mercury in the Western Hemisphere, the New Idria Mine was undeniably important in California's history.[12] The late 1960s had been a particularly flush time for the New Idria Company, because of a quicksilver price boom (due in part to demand for mercury during the Vietnam War), and the mine and town of New Idria prospered.[13] But the New Idria Company's success ended in the early 1970s, doomed by environmental regulations that ended the active mining of cinnabar and its reduction into mercury in the United States. With no prospect of mercury mining ever resuming in the United States, the New Idria Company faded and the town and mine buildings of New Idria were sold in 1973, ending over 120 years of continuous operation, by far the longest active period of any mine in California history.

Today the New Idria Mine, like the many mercury mines of California and the western United States, is in ruins (Figure 7.6). In California the Coast Ranges from Santa Barbara north to the Oregon border are littered with the remains of nearly forty major mercury mine sites and hundreds of minor prospects. Some of these former mines appear as place-names on old maps; most sites, however, are known only to locals and a few historians and geologists. These sites, barely accessible via eroded mining roads, are visible in the landscape as scarred earth and piles of mine tailings, accented with decaying structures and odd remnants of machinery not worthy of salvage (Figures 7.7 and 7.8). Old mercury mines, such as New Idria, are part of the lasting legacy of the quicksilver industry in the landscapes of California and the West. Beyond the sites of mercury production is the environmental legacy of mercury—

FIGURE 7.6 Images of the New Idria Reduction Plant and Miners' Houses, 2002

The reduction plant (top) featured rotary kiln reduction technology and was built in the early 1920s. A son and a brother of Herbert Hoover, himself a mining engineer by training, were part owners of New Idria in the late 1930s, when this technology was state-of-the-art for cinnabar reduction. The miners' houses (bottom), simple board-and-batten structures, show change over time through the construction of additions. (Photographs by author.)

FIGURE 7.7 The Corona Mine Four-Shaft Scott Furnace, Napa County

The Corona Mine Scott furnace was built with the standard brick-and-tile interior, but was faced with a local stone. Scott furnaces required "seasoning," that is, running ore until the bricks and tiles on the interior became saturated with mercury. After saturation, the furnace would produce the mercury it should, based on the richness of the ore. Today the interior brick and tile have long been removed, processed for their mercury content. However, the stone facing remains, showing the scale of these furnaces. (Top photograph: Bradley, *Quicksilver Resources of California Bulletin No. 78*, 108.) The construction photograph shows the Corona furnace being built. (Bottom: contemporary photograph by author, 2002.)

FIGURE 7.8 The Buckeye Mine Scott Furnace, Sulphur Creek District, 2002

This two-shaft Scott furnace was reputedly never used, explaining why it was not torn apart and the bricks processed for their mercury content. The photograph shows brick ledges for tiles remaining at the center of the furnace. (Photograph by author.)

from polluted streams, lakes, and bays; to toxic sites of gold and silver mining; to the concentrations of mercury in the food we eat.

Beyond concerns about mercury pollution in the environment, Californians have largely forgotten the quicksilver mining industry and ignored its legacy. This is the case despite substantial marks on the California and western landscape. The legacy of mercury mining is around us, if only we choose to see it, study it, and learn from it. New Almaden is the most visible site, with a county-level park and a smart but modest museum. New Almaden does not, however, command the respect it deserves. Few people have heard of it; even fewer ever visit despite a remarkably intact village. However, New Almaden is remarkable: it was both the first mine and the single richest mine in the state; it was the most significant colonial hacienda plantation in California; its riches defeated the powerful Rothschild cartel in its goal of a worldwide mercury monopoly for many years; it was an important engine behind the Bank Crowd, especially in their Comstock dealings; and most important, perhaps, it gave the gold and silver mining industries of the American West a domestic source of quicksilver. Without this quicksilver, the history of the West would have been very different. New Almaden embodies the global connections of early California, including connections to the ascendant British Empire and the remnants of the dying Spanish Empire. The story of New Almaden includes both California's colonial past and, for many immigrant groups, a narrative of becoming American.

Despite its significance, New Almaden remains largely unknown. It and the other remnants of the quicksilver industry are largely forgotten because they are not the embodiment of the dominant myths of California that come to us from the focus on gold and silver. They do not tell the stories Californians want to tell about themselves or the stories people want to tell about us. Eustace Barron and Alexander Forbes have not been embraced as founding fathers of California. Californians do not wish to remember the colonial past represented by Hacienda Nuevo Almaden. The history of our hierarchical ordering by race, ethnicity, and class is complex and messy, a hard story to tell in California today. The long history of industrial capital challenges our myth of the rugged individual. New Almaden is not only a landscape of the 1840s; by looking at New Almaden, we can learn authentic stories about ourselves.

Quicksilver in California tells us a story different from that told by gold and silver. A focus on mercury opens new understandings of the state and of ourselves. For instance, within the quicksilver industry no worker ever struck it rich. There is no free-labor ideology to be found in the industry; wage labor was a way of life, not a stage on the path to higher development. The founding of a quicksilver industry in California begins with the story of European colonialism. With the quicksilver industry, we can see clearly how the tendrils of colonialism work down through time into the business organization and culture of nineteenth-century California. We can see within the industry how the colonial hacienda continued down through the decades as a spatial ordering of California landscapes. Through quicksilver we gain a new perspective on the workings of social and cultural power in the formation of California, a perspective that illuminates much in California and western history that is left unexamined when we focus myopically on gold and silver.

Notes

Portions of this chapter were originally published in *Journal of the West* 43, no. 1 (Winter 2004); © Journal of the West 2004, reprinted with permission of ABC-CLIO.

1. 1 F. 481 (C.C.D. Cal. 1880).

2. Warrant, *The People v. Tiburcio Parrott*, City and County of San Francisco, February 20, 1880; National Archive and Records Administration, San Bruno, California.

3. The ore began a few yards beneath the surface, and soon became very rich. The surface layer was white silica and sulphur, which was dusty and produced great glare. Beneath this surface layer, the rock was intensely crushed and irregularly metamorphosed. Beneath this next layer, the basalt was shattered. The Sulphur Bank Mine was probably the only mine in California where active deposition of cinnabar was taking place during mining. Cinnabar was being deposited by the scalding sulphurous waters that were being pushed up through and saturating the various strata of rock. This led the mine to become possibly the most noxious and intensely difficult mine in the state.

4. When the Parrott group took over the mine, it already had two noncontinuous furnace types. By 1877, reports indicate that the mine had five "Lightner" furnaces (named after and supposedly developed by the superintendent), a Knox-Osborn furnace from the Redington Mine design, additional banks of retorts, and extensive works for drying out the ore, a major problem in a thermally active zone.

5. Goss, *Gold and Cinnabar*.

6. *Independent Calistogan*, May 18, 1880.

7. Miners' Association of California, "The Debris Question."

8. Ibid., 3.

9. Cyanide processing would greatly curtail the demand for mercury for mining markets by the early 1890s. See Spude, "Cyanide and the Flood of Gold."

10. For information on the Texas quicksilver industry, see Andrew Gomez, *Quicksilver*; Gomez, *A Most Singular Country*; Tyler, *The Big Bend*; and Library of Congress, Prints and Photographs Division, Historic American Engineering Record, National Park Service, Mariscal Mine Recording Project, 1997.

11. From the flyer for the sale by Milton J. Wershow Co., Auctioneers and Realtors.

12. The mine proved to have vast reserves of low-grade ore, and from the 1920s into the early 1970s the New Idria Company dominated quicksilver production in the country.

13. Mercury fulminate was the crucial ingredient in bomb detonators.

Manuscript Collections and Archives

BANCROFT LIBRARY, UNIVERSITY OF CALIFORNIA, BERKELEY
Babcock Collection.
Josiah Belden Papers, 1832–1903.
James Alexander Forbes Papers, 1845–53.
Guadalupe Mine Collection, 1854–1921.
Hearst Mining Collection of Views by C. E. Watkins.
Robert B. Honeyman Jr. Collection.
J. B. Randol. Papers Relating to Quicksilver Mining, CA, 1849–1894.
Irving Murray Scott Correspondence, November 24, 1871–March 1, 1882.

CALIFORNIA STATE LIBRARY, SACRAMENTO
Kenneth Rank–New Idria Mines Collection, 1867–1973.
New Idria Mining Company Records, 1854–67.
New Idria Quicksilver Mining Company Collection #2.

HISTORY SAN JOSE RESEARCH LIBRARY
Historic Images of the New Almaden Quicksilver Mines.
Perham History Files, 1800–1980.
The Jimmy Schneider and Robert Bulmore Collection of New Almaden Mine Materials Collection, no. 1978-251.

LIBRARY OF CONGRESS
Photographs Division, Historic American Engineering Record, National Park Service.

NATIONAL ARCHIVES AND RECORDS ADMINISTRATION
San Bruno, California.

STANFORD UNIVERSITY
New Almaden Mine Collection, 1845–1973.

Manuscript Documents

Bancroft, Hubert Howe, Stephen Franklin, Andrew J. Ralston, Edney S. Tibbey, George H. Morrison, Thomas Bell, Andrew B. Forbes, and Collection Hubert Howe Bancroft. *Biography*

Bibliography

of William C. Ralston Prepared for Chronicles of the Kings: And Material Used in Its Preparation, 1886–1889*. San Francisco: Bancroft Library.
Bell, Thomas Frederick. "Dictation Concerning William Ralston." In *H. H. Bancroft—Chronicles of the Kings*. San Francisco: Bancroft Library.
Bell, Thomas Frederick, and William Henry Long. "Dictation Concerning the New Almaden Quicksilver Mines: Ms., June 15, 1886." In Hubert Howe Bancroft Collection. San Francisco: Bancroft Library.
Disturnell, Nathaniel. "Diary Relating to Mercury Mining." Wilbur Springs, CA, 1874–76. San Francisco: Bancroft Library, MSS C-F, 196.
Empire Consolidated Mines. "Central Consolidated Quicksilver Mines U.S. Lot 3605, Empire Consolidated Quicksilver Mines U.S. Lot 3606." Bancroft Library, University of California, Berkeley, Bancroft Microfiche 2479.
Integral Quicksilver Mining Company. *Prospectus of Integral Quicksilver Mining Company*. 1891.
Rawson, A. W. *A. W. Rawson Diary*, 1856. Bancroft Library, University of California, Berkeley.
Watkins, Carleton E., E. A. Day, and C. F. Richardson. *Views of New Almaden*. Bancroft Library, University of California, Berkeley, 186?.
Wilder, James William, William Casburn, Donald L. Gustafson, Eleanor Swent, Bancroft Library Regional Oral History Office, and Knoxville District / McLaughlin Mine Project. *Owner of the Shot Mining Company, Manhattan Mercury Mine, 1965–1981: Oral History Transcript 1996*. San Francisco: Bancroft Library.
Wood, R. E. *Store and Office, R. Q. Mining Co., Knoxville, Napa County, Cal.* San Francisco: Bancroft Library.

California Newspapers

Daily Alta California
Independent Calistogan
Lower Lake Bulletin
Mining and Scientific Press
Morning Call
Napa Register
Russian River Flag
San Francisco Bulletin
San Francisco Call
San Francisco Examiner
San Jose Mercury
Sonoma Democrat

Other Sources

Abu-Lughod, Janet L. *Before European Hegemony: The World System, A.D. 1250–1350*. New York: Oxford University Press, 1989.
Abu-Lughod, Janet L. *New York, Chicago, Los Angeles: America's Global Cities*. Minneapolis: University of Minnesota Press, 1999.
Allen, John, Doreen B. Massey, and Allan Cochrane. *Rethinking the Region*. New York: Routledge, 1998.
Almaguer, Tomas. *Racial Fault Lines: The Historical Origins of White Supremacy in California*. Berkeley: University of California Press, 1994.
Alt, David, and Donald W. Hyndman. *Roadside Geology of Northern and Central California*. Missoula, MT: Mountain Press, 2000.
American Quicksilver Company. *American Quicksilver Company, of California: Prospectus*. New York: American Quicksilver Company, 1849.
Andrews, Kenneth R. *The Spanish Caribbean: Trade and Plunder, 1530–1630*. New Haven: Yale University Press, 1978.
Ascher, Leonard. *Lincoln's Administration and the New Almaden Scandal*. California, 1936. http://dx.doi.org/10.2307/3633317.
Ascher, Leonard William. "The Economic History of the New Almaden Quicksilver Mine, 1845–1863." Ph.D. diss., University of California Berkeley, 1934.
Ashburner, William, James D. Hague, and Thomas Price. *Report of William Ashburner, James D. Hague, Thomas Price, M.C. Vincent: On the Properties of the Sulphur Bank Quicksilver Mining Co*. San Francisco: Edward Bosqui, 1876.
Aubury, Lewis E. *The Quicksilver Resources of California*. Ed. California State Mining Bureau. *Bulletin No. 27*. Sacramento: California State Printing Office, 1903.
Bakewell, Peter J. *Miners of the Red Mountain: Indian Labor in Potosí, 1545–1650*. Albuquerque: University of New Mexico Press, 1984.
Bakewell, Peter J., ed. *Mines of Silver and Gold in the Americas*. Brookfield, VT: Ashgate, 1997.
Bakewell, Peter J. *Silver and Entrepreneurship in Seventeenth-Century Potosí: The Life and Times of Antonio Lopez de Quiroga*. Albuquerque: University of New Mexico Press, 1988.
Bakewell, Peter J. *Silver Mining and Society in Colonial Mexico: Zacatecas, 1546–1700*. Cambridge: Cambridge University Press, 1971. http://dx.doi.org/10.1017/CBO9780511572692.
Barker, Leo R., and Ann E. Huston, eds. *Death Valley to Deadwood; Kennecott to Cripple Creek: Proceedings of the Historic Mining Conference, January 23–27, 1989, Death Valley National Monument*. San Francisco: National Park Service, 1990.
Barron, Eustace W. *Algunas declaraciones en el asunto de Nuevo Almaden en la Alta California*. Mexico City: Imprenta de Andrade y Escalante, 1859.
Bartlett, John Russell. *Personal Narrative of Explorations and Incidents in Texas, New Mexico, California, Sonora, and Chihuahua: A Rio Grande Classic*. Chicago: Rio Grande Press, 1966.
Basso, Matthew, Laura McCall, and Dee Garceau. *Across the Great Divide: Cultures of Manhood in the American West*. New York: Routledge, 2001.

Beck, Charlotte. *Models for the Millennium: Great Basin Anthropology Today*. Salt Lake City: University of Utah Press, 1999.

Becker, George F. *Atlas to Accompany a Monograph on the Geology of the Quicksilver Deposits of the Pacific Slope*. Washington, DC: U.S. Geological Survey, 1887.

Becker, George F. *Geology of the Quicksilver Deposits of the Pacific Slope*. Monographs of the United States Geological Survey, vol. 13. Washington, DC: Government Printing Office, 1888.

Berger, Alan. *Reclaiming the American West*. New York: Princeton Architectural Press, 2002.

Berman, Marshall. *All That Is Solid Melts into Air: The Experience of Modernity*. New York: Simon and Schuster, 1982.

Billings, Frederick. *Letters from Mexico*. Woodstock, VT: Elm Tree Press, 1936.

Boulland, Michael, and Arthur Boudreault. *Images of New Almaden*. Charleston, SC: Arcadia Publishing, 2006.

Bradley, Walter W. *Quicksilver Resources of California Bulletin No. 78*. Sacramento: California State Mining Bureau, California State Printing Office, 1918.

Braudel, Fernand. *Civilization and Capitalism, 15th–18th Century: The Perspectives of the World*. London: Fontana Press, 1984.

Braudel, Fernand. *The Mediterranean and the Mediterranean World in the Age of Philip II*. Vol. 1. New York: Harper and Row, 1972.

Brechin, Gray. *Imperial San Francisco*. Berkeley: University of California Press, 1999.

Brewer, William H. *Up and Down California in 1860–1864*. Ed. Francis P. Farquhar. New Haven: Yale University Press, 1930.

Browne, D. Mackenzie, ed. *China Trade Days in California: Selected Letters from the Thompson Papers, 1832–1863*. Berkeley: University of California Press, 1947.

Browne, J. Ross. "Down in the Cinnabar Mines: A Visit to New Almaden in 1865." *Harper's New Monthly Magazine* 31, no. 185 (1865): 545–60.

Buckley, James M. "Building the Redwood Region: The Redwood Lumber Industry and the Landscape of Northern California, 1850–1929." Ph.D. diss., University of California, Berkeley, 2000.

Buel, Eustace Barron, William Forbes, and Alexander Forbes. *Eustace Barron and William and Alex. Forbes*. Washington, DC: GPO, 1850.

Bulmore, Robert R., S. E. Winn, and J. Ross Browne. *Views of New Almaden and the New Almaden Quicksilver Mines*. Bancroft collection, n.d., Banc Pic 19xx 116-Album. Graphic, 1870.

Butler, Phyllis F. "New Almaden's Casa Grande." *California Historical Quarterly* 54, no. 4 (Winter): 315–22.

Byington, Margaret F. *Homestead: The Households of a Mill Town*. Pittsburgh: University of Pittsburgh Press, 1974.

Cahill, Thomas. *How the Irish Saved Civilization: The Untold Story of Ireland's Heroic Role from the Fall of Rome to the Rise of Medieval Europe*. New York: Anchor Books, 1995.

California Division of Mines and Geology. *Geologic Map and Sections of the Mayacamas Quicksilver District: Sonoma, Lake and Napa Counties, California*. Sacramento: Division of Mines, 1946.

California Legislature. *Joint Resolution of the Legislature of California, Relating to the New Almaden Mine*. Washington, DC: Gideon Printer, 1860.

California Picacho Quicksilver Company. *Prospectus of the California Picacho Quicksilver Company, San Benito Co., California: Incorporated under the Laws of the State of New York*. New York: The Company, 1879.

California Quicksilver Mining Co. *Company Store Day Book, August 1, 1874–May 25, 1875*. Yolo Company, CA.

Candee, Richard M. "New Towns of the Early New England Textile Industry." In *Perspectives in Vernacular Architecture*, ed. Camille Wells, 31–50. Vernacular Architecture Forum, 1982. http://dx.doi.org/10.2307/3514265.

Carter, Thomas, ed. *Images of an American Land*. Albuquerque: University of New Mexico Press, 1997.

Castillero, Andrés, and United States District Court (California: Northern District). *The United States vs. Andres Castillero: "New Almaden" Transcript of the Record*. 4 vols. San Francisco, 1859.

Chan, Sucheng. *Occupational Structure and Social Stratification in Chinese Immigrant Communities in Nineteenth-Century Rural California*. Paper presented at the Ninety-sixth Annual Meeting of the American Historical Association, Los Angeles, December 28–30, 1981.

Chen, Thomas. *Chinese San Francisco*. Palo Alto, CA: Stanford University Press, 2000.

Chicago Quicksilver Mining Company. *Chicago Quicksilver Mining Co. Prospectus*. San Jose.

Chinn, Thomas W., H. Mark Lai, Philip P. Choy, and Chinese Historical Society of America. *A History of the Chinese in California: A Syllabus*. San Francisco: Chinese Historical Society of America, 1973.

Christy, Samuel Benedict. "On the Genesis of Cinnabar Deposits." *American Journal of Science* 3, no. 17 (1878): 453–63.

Christy, Samuel Benedict. *Quicksilver-Condensation at New Almaden, California: A Paper Read before the American Institute of Mining Engineers, at the Chattanooga Meeting, May, 1885*. Rev. ed. Philadelphia: The Institute, 1885.

Christy, Samuel Benedict. *Quicksilver Reduction at New Almaden*. Philadelphia: Sherman & Co., 1884.

Clark, Edward. *Quicksilver Mining: New Hypothesis Regarding the Origin and Formation of Mineral Deposits, as Applied to Elucidate the Quicksilver Lodes of California*. Napa City, CA: Napa Register, 1872.

Coates, Peter. "Garden and Mine, Paradise and Purgatory: Landscapes of Leisure and Labor in California." In *"Nature's Nation" Revisited*, ed. Hans Bak and Walter W. Holbling, 147–67. Amsterdam: Vrije Universiteit Press, 2003.

Cobb, Gwendolin B. *Potosí and Huancavelica: Economic Bases of Peru, 1545–1640*. University of California, Berkeley, 1947.

Coignet, M. *Rapport sur les mines de New-Almaden (Californie)*. Paris: Dunod, 1866.

Coleman. *Field Trip Guidebook to the New Idria Area*. Palo Alto: Stanford University, 1986.

Commercial Steam Book and Job Presses (San Francisco, Calif.). *Correspondence in Relation to the New Almaden Quicksilver Mine of California, between the Counsel for the Proprietors and the Government*. San Francisco: Commercial Steam Presses, 1859.

Commercial Steam Book and Job Presses. *Further Correspondence in Relation to the New Almaden Quicksilver Mine of California: Between the Counsel for the Proprietors and the Government*. San Francisco: Commercial Steam Presses, 1859.

Coomes, Mary L. "From Pooyi to the New Almaden Mercury Mine: Cinnabar, Economics, and Culture in California to 1920." Ph.D. diss., University of Michigan, 1999.

Corcoran, May S., and Helen Bolton. *The Romance of Quicksilver*. Oakland, CA, 1922.

Crawford, Margaret. *Building the Workingman's Paradise*. New York: Verso, 1995.

Cronon, William. *Changes in the Land*. New York: Hill and Wang, 1983.

Cronon, William. *Nature's Metropolis: Chicago and the Great West*. New York: W. W. Norton, 1991.

Cross, Ira B. *Financing an Empire: History of Banking in California*. Vol. 1. Chicago, San Francisco, Los Angeles: The S. J. Clarke Publishing Co., 1927.

Day, David T. *Mineral Resources of the United States*. Washington, DC: Government Printing Office, 1892.

del Castillo, Antonio. *Memoria sobre las minas de Azogue de América: Conteniendo el resumen de los reconocimientos practicados en las de México, y la descripción de las de Alta California y Huancavelica*. Mexico City: Impr. de I. Escalante y Compañía, 1871.

Dickens, Charles, ed. *All the Year Round* 13, no. 318 (May 27, 1865): 424–28.

Doble, John. *John Doble's Journal and Letters from the Mines*. Ed. Charles L. Camp. Volcano, CA: Volcano Press, 1999.

Doti, Lynne Pierson. *Banking in an Unregulated Environment: California, 1878–1905*. New York: Garland Publishing, 1995.

Doti, Lynne Pierson, and Larry Schweikart. *Banking in the American West: From the Gold Rush to Reregulation*. Norman: University of Oklahoma Press, 1991.

Downer, S. A. "On Her Trip into the New Almaden Mine." *The Pioneer; or California Monthly Magazine* 2 (1854): 220–28.

Doyle, John T. *A Letter to the President of the United States, in Reply to the Attorney General's Report of the Resolution of the Legislature of the State of California*, 1860.

Duschak, L. H., and Curt N. Schuette. "The Metallugy of Quicksilver," *Bureau of Mines Bulletin* 222. Washington, DC: U.S. Government Printing Office, 1925.

Eckel, Edwin Butt, Robert G. Yates, and A. E. Granger. *Quicksilver Deposits in San Luis Obispo County and Southwestern Monterey County, California*. Washington, DC: U.S. Government Printing Office, 1941.

Egenhoff, Elisabeth L. *De Argento Vivo: Historic Documents on Quicksilver and Its Recovery in California Prior to 1860.* San Francisco: California Division of Mines, 1953.

Egleston, T. "Notes on the Treatment of Mercury in North California." *Transactions of the American Institute of Mining Engineers* 3, 1875.

Ehrenberg, Richard. *Capital and Finance in the Age of the Renaissance: A Study of the Fuggers and Their Connections.* Reprints of Economic Classics. New York: Augustus M. Kelley, 1963.

Emmons, David M. *The Butte Irish: Class and Ethnicity in an American Mining Town, 1875–1925.* Urbana: University of Illinois Press, 1989.

Everhart, Donald L. "Quicksilver Deposits at the Sulphur Bank Mine, Lake County, California." In *Report XLII of State Mineralogist*, 29 pp. Sacramento: State of California, 1946.

Eysen, Edward, and Sunlight Printing. *Plat of the Property of the Quicksilver Mining Co., New Almaden, Cal.* San Francisco: Sunlight Printing, 1888.

Felton, Charles Norton. "The Tariff Speech of Hon. Charles N. Felton, of California." Speech presented in the House of Representatives, Washington, DC, May 17, 1888.

Ferguson, Niall. *The House of Rothschild: The World's Banker, 1849–1999.* New York: Viking, 1999.

Flynn, Dennis O. *World Silver and Monetary History in the 16th and 17th Centuries.* Brookfield, VT: Ashgate, 1996.

Flynn, Dennis O., Lionel Frost, and A.J.H. Latham, eds. *Pacific Centuries: Pacific and Pacific Rim History since the Sixteenth Century.* London and New York: Routledge, 1999. http://dx.doi.org/10.4324/9780203445662.

Flynn, Dennis O., A.J.H. Latham, and Sally M. Miller, eds. *Studies in the Economic History of the Pacific Rim.* London: Routledge, 1998. http://dx.doi.org/10.4324/9780203279908.

Foote, Mary Hallock. *New Almaden: Or, a California Mining Camp: Life in 1877 at New Almaden as Pictured in Word and Illustration.* Fresno, CA: Valley Publishers, 1969.

Foote, Mary Hallock. *A Victorian Gentlewoman in the Far West.* Ed. Rodman Paul. San Marino, CA: Henry E. Huntington Library and Art Gallery, 1972.

Forbes, Alexander. *California: A History of Upper and Lower California from Their First Discovery to the Present Time.* San Francisco: J. H. Nash, 1937 [1839].

Forbes, James Alexander, and Andrés Castillero. *United States. In the Claim of Andres Castillero to the Mine of New Almaden: Deposition of James Alexander Forbes, a Witness Called by the United States, on the 14th of July, 1858.* San Francisco: Commercial Book and Job Steam Printing Establishment, 1858.

Forbes, James Alexander, and United States District Court (California: Northern District). *The United States vs. Andres Castillero: San Francisco, July 16th, 1858.* San Francisco, 1858.

Forstner, William. *The Quicksilver Resources of California.* Sacramento: W. W. Shannon Superintendent of State Printing, 1903.

Forstner, William. *The Quicksilver Resources of California.* Sacramento: W. W. Shannon Superintendent of State Printing, 1908.

Fossatt, Charles, and Quicksilver Mining Company. *Facts Concerning the Quicksilver Mines in Santa Clara County, California, for Private Circulation*. New York: R. C. Root, Anthony, 1859.

Francaviglia, Richard. *Hard Places*. Iowa City: University of Iowa Press, 1991.

Frusetta, Peter C. *Quicksilver Country: California's New Idria Mining District*. Tres Pinos, CA: Peter Frusetta, 1991.

Geological Survey of California, and J. D. Whitney. *Geology*. Philadelphia: Caxton Press of Sherman & Co., 1865.

Gillenkirk, Jeff, and James Motlow. *Bitter Melon: Stories from the Last Rural Chinese Town in America*. Seattle: University of Washington Press, 1987.

Goldwater, Leonard J. *Mercury: A History of Quicksilver*. Baltimore: York Press, 1972.

Gomez, Arthur R. *A Most Singular Country: A History of Occupation in the Big Bend*. Salt Lake City: Brigham Young University Press, 1990.

Goss, Helen Rocca. *Gold and Cinnabar: The Life of Andrew Rocca, California Pioneer*. Montgomery, AL: H. R. Goss, 1990.

Goss, Helen Rocca. *The Life and Death of a Quicksilver Mine*. Los Angeles: Historical Society of Southern California, 1958.

Greene, Mott. *Geology in the Nineteenth Century*. Ithaca: Cornell University Press, 1983.

Greever, William S. *Bonanza West*. Moscow: University of Idaho Press, 1963.

Gregory, Derek. *Geographical Imaginations*. Cambridge, MA: Blackwell Publishers, 1994.

Guadalupe Mining Company, Santa Clara Mining Association of Baltimore, Davy Mining and Investment Company, D & C Mining Company, Century Mining Company, New Guadalupe Mining Company, and Spanish Springs Mining Company, Tonopah, Nevada. *Guadalupe Mine Records, 1854–1921*.

Haas, Lisbeth. *Conquests and Historical Identities in California*. Berkeley: University of California Press, 1995.

Hall, Jacquelyn, James Leloudis, Robert Korstad, Mary Murphy, LuAnn Jones, and Christopher B. Daly. *Like a Family: The Making of a Southern Cotton Mill Town*. New York: W. W. Norton, 1987. http://dx.doi.org/10.5149/uncp/9780807848791.

Hammond, George P., ed. *Thomas O. Larkin Papers*. Berkeley: University of California Press, 1968.

Hamnett, Brian R. *Politics and Trade in Southern Mexico, 1750–1821*. Cambridge: Cambridge University Press, 1971.

Hardesty, Donald L. "Archaeological Models of the Modern World in the Great Basin: World Systems and Beyond." In *Current Models in Great Basin Anthropology*, ed. C. Beech, 213–19. Salt Lake City: University of Utah Press, 1999.

Hardesty, Donald L. *The Archaeology of Mining and Miners: A View from the Silver State*. Pleasant Hill, CA: Society for Historical Archaeology, 1988.

Hardesty, Donald L. "Class, Gender Strategies, and Material Culture in the Mining West." In *Those of Little Note: Gender, Race, and Class in Historical Archaeology*, ed. E. Scott, 129–49. Tucson: University of Arizona Press, 1994.

Hardesty, Donald L. "Power and the Industrial Mining Community in the American West." In *Social Approaches to an Industrial Past. The Archaeology and Anthropology of Mining*, ed. A. Bernard Knapp, Vincent C. Pigott, and Eugenia W. Herbert, 81–96. New York: Routledge, 1998.

Harlow, Neal. *California Conquered*. Berkeley: University of California Press, 1982.

Harte, Bret. *Story of a Mine*. Authorized ed. London: George Routledge & Sons, 1877.

Harvey, David. *The Condition of Postmodernity: An Enquiry into the Origins of Cultural Change*. Oxford: Blackwell, 1989.

Harvey, David. *Consciousness and the Urban Experience: Studies in the History and Theory of Capitalist Urbanization*. Baltimore: Johns Hopkins University Press, 1985.

Harvey, David. *The Urban Experience*. Oxford: B. Blackwell, 1989.

Haus, Donald O. *The Knoxville-Redington Mine, 1860 to 1882: Its Development and Place in the California Quicksilver Industry*. Sacramento: Sacramento State College, 1965.

Heath, Hilarie J. "British Merchant Houses in Mexico, 1821–1860: Conforming Business Practices and Ethics." *Hispanic American Historical Review* 73, no. 2 (May 1993): 261–90.

Holdredge, Helen. *Mammy Pleasant's Partner*. New York: G. P. Putnam's Sons, 1954.

Holliday, J. S. *Rush for Riches*. Berkeley: University of California Press, 1999.

Holliday, J. S. *The World Rushed In*. New York: Simon and Schuster, 1981.

Howard, Arthur D. *Geologic History of Middle California*. Berkeley: University of California Press, 1979.

Howser, Huell, Home Savings of America, and KCET (Los Angeles). *California's Gold: Quicksilver*. Video recording. Los Angeles: Huell Howser Productions, 1998.

Hudson, Lynn M. *The Making of "Mammy Pleasant": A Black Entrepreneur in Nineteenth-Century San Francisco*. Urbana and Chicago: University of Illinois Press, 2003.

Hutchings, James M. *Scenes of Wonder and Curiosity in California*. New York and San Francisco: A. Roman & Co., 1870.

Ignatiev, Noel. *How the Irish Became White*. New York: Routledge, 1995.

Isaac, Rhys. *The Transformation of Virginia, 1740–1790*. Chapel Hill: University of North Carolina Press, 1999.

Isenberg, Andrew. *Mining California: An Ecological History*. New York: Hill & Wang, 2005.

James, Ronald M. *The Roar and the Silence: A History of Virginia City and the Comstock Lode*. Reno: University of Nevada Press, 1998.

James, Ronald M., and C. Elizabeth Raymond, eds. *Comstock Women: The Making of a Mining Community*. Reno: University of Nevada Press, 1998.

Jenkin, A. K. Hamilton. *The Cornish Miner*. London: George Allen & Unwin Ltd., 1927.

Jennings, Hennen. *The Quicksilver Mines of Almaden and New Almaden: Comparative View of Their Extent, Production, Costs of Work, Etc*. San Francisco: Bacon & Company Excelsior Press, 1886.

Johnson, Fremont T., and Spangler Ricker. *Investigation of Oat Hill Mercury Mine, Napa County, Calif*. Washington, DC: U.S. Bureau of Mines, 1949.

Johnson, Kenneth M. *The New Almaden Quicksilver Mine: With an Account of the Land Claims Involving the Mine and Its Role in California History.* Georgetown, CA: Talisman Press, 1963.

Johnson, Reverdy, Jeremiah S. Black, and Charles Fossatt. *The United States vs. Charles Fossatt: Motion to Dismiss Appeal. Brief for Appellants.* Washington, DC, 1858.

Johnson, Susan Lee. *Roaring Camp: The Social World of the California Gold Rush.* New York: W. W. Norton, 2000.

Johnston, Andrew S. "Quicksilver Landscapes: Mercury Mining and Chinese Labor in Northern California to 1880." *Journal of the West* 43, no. 1 (Spring 2004): 21–29.

Johnston, Marc W. "Faith with Our Victims: the Litigation of Mexican Land Grants in California after the American Accession—A Case Study of the New Almaden Claims." M.A. thesis, Harvard College, Massachusetts, 1974.

Jones, Idwal. *Vermilion.* New York: Prentice-Hall, 1947.

King, Anthony D., ed. *Culture, Globalization and the World System: Contemporary Conditions for the Representation of Identity.* Minneapolis: University of Minnesota Press, 1977.

Klubock, Thomas Miller. *Contested Communities: Class, Gender, and Politics in Chile's El Teniente Copper Mine, 1904–1951.* Durham, NC: Duke University Press, 1998.

Knapp, Arthur Bernard, Vincent C. Pigott, and Eugenia W. Herbert. *Social Approaches to an Industrial Past: The Archaeology and Anthropology of Mining.* London: Routledge, 1998.

Konrad, Herman. *A Jesuit Hacienda in Colonial Mexico: Santa Lucia, 1576–1767.* Stanford, CA: Stanford University Press, 1980.

Kuss, M. H. *Mémoire sur les mines et usines d'Almaden.* Paris: Dunod, 1878.

Lai, H. Mark, Joe C. Huang, Don Wong, and Chinese Culture Foundation. *The Chinese of America, 1785–1980: An Illustrated History and Catalog of the Exhibition.* San Francisco: Chinese Culture Foundation, 1980.

Lake Valley Quicksilver Mining Company of New York. *The Lake Valley Quicksilver Mining Co. of New York. Capital Stock, $10,000,000 Divided into 100,000 Shares, Each $100. Also the Reports by Profs. Spence, Silversmith.* New York: Printed at the Office of the "Pacific Index," 1865.

Lanyon, Milton, and Laurence Bulmore. *Cinnabar Hills: The Quicksilver Days of New Almaden.* Los Gatos, California: Village Printers, 1982.

Larkin, Thomas Oliver, Anna Marie Hager, George Peter Hammond, Everett Gordon Hager, and Bancroft Library. *The Larkin Papers: Personal, Business, and Official Correspondence of Thomas Oliver Larkin, Merchant and United States Consul in California.* Berkeley: Published for the Bancroft Library by the University of California Press, 1951.

Layres, Augustus. *Other Side of the Chinese Question; to the People of the United States & the Honorable Senate & House of Representatives; Testimony of California's Leading Citizens; Read & Judge, Chinese Immigration Pamphlets;* Vol. 3, No. 11. San Francisco, 1886.

Lee, Lai To. *Early Chinese Immigrant Societies: Case Studies from North America and British Southeast Asia.* Asian Studies Series. Singapore: Heinemann Asia, 1988.

Lee, Rose Hum. "The Growth and Decline of Chinese Communities in the Rocky Mountain Region." Ph D. diss., University of Chicago, 1947.

Limerick, Patricia Nelson. "Disorientation and Reorientation: The American Landscape Discovered from the West." *Journal of American History* 79, no. 3 (1992): 1021–49. http://dx.doi.org/10.2307/2080797.

Limerick, Patricia Nelson. *The Legacy of Conquest: The Unbroken Past of the American West.* New York: W. W. Norton, 1987.

Limerick, Patricia Nelson, Clyde A. Milner II, and Charles E. Rankin. *Trails: Toward a New Western History.* Lawrence: University Press of Kansas, 1991.

Lingenfelter, Richard E. *The Hardrock Miners.* Berkeley: University of California Press, 1974.

Livermore, Horatio. *Annual Report of the Redington Mine.* San Francisco: s.n., 1876.

Lohmann, Villena. *Guillermo: Las minas de Huancavelica en los siglos XVI y XVII.* 2. Lima: Pontificia Universidad Católica del Perú Fondo Editorial, 1999.

Lord, John Keast. *The Naturalist in British Columbia.* Vol. 1. London: Richard Bentley, 1866.

Lyman, C. S. *On Mines of Cinnabar in Upper California: The American Journal of Science and Arts.* New Haven: B. L. Hamlen, Printer to Yale College, 1848.

Lyman, Chester Smith, and United States District Court (California: Northern District). "Mapa del terreno de la Mina de Nueva Almaden." 1848.

Malone, Michael P. "Beyond the Last Frontier: Toward a New Approach to Western American History." In *Trails: Toward a New Western History,* ed. Patricia Nelson Limerick, Clyde A. Milner II, and Charles Rankin, 139–60. Lawrence: University Press of Kansas, 1991.

Mason, Jesse D. *History of Santa Barbara and Ventura Counties.* Berkeley: Howell and North, 1961.

Massey, Doreen B. *Space, Place, and Gender.* Minneapolis: University of Minnesota Press, 1994.

Mayo, John. *Commerce and Contraband on Mexico's West Coast in the Era of Barron, Forbes & Co., 1821–1859.* New York: Peter Lang, 2006.

McAllister, Matthew Hall, and Andrés Castillero. *Andres Castillero vs. The United States, on Appeal from the Decree of the United States Commissioners to Ascertain and Settle the Private Land Claims in the State of California.* Brief, 1859.

McCarl, Robert. *Contested Space.* Salt Lake City: University of Utah Press, 1997.

McKinney, Gage, and New Almaden County Quicksilver Park Association. *A High and Holy Place: A Mining Camp Church at New Almaden.* New Almaden, CA: New Almaden County Quicksilver Park Association, 1997.

McWilliams, Carey. *California: The Great Exception.* Berkeley: University of California Press, 1999.

Mellinger, Philip J. *Race and Labor in Western Copper: The Fight for Equality, 1896–1918.* Tucson: University of Arizona Press, 1995.

Mercier, Laurie. *Anaconda: Labor, Community, and Culture in Montana's Smelter City.* Urbana: University of Illinois Press, 2001.

Mills, Sara. *Discourses of Difference: An Analysis of Women's Travel Writing and Colonialism.* London: Routledge, 1993.

Miners' Association of California. "The Debris Question." Alta Printing House, August 1, 1881.

Moodie, T. Dunbar, and Vivienne Ndatshe. *Going for Gold: Men, Mines, and Migration*. Berkeley: University of California Press, 1994.

Muckler, Billie Phyllis. "History of the Guadalupe Quicksilver Mine." M.A. thesis, University of California, 1941.

Murphy, Mary. *Mining Cultures: Men, Women, and Leisure in Butte, 1914–1941*. Urbana: University of Illinois Press, 1997.

Nash, June C. *We Eat the Mines and the Mines Eat Us: Dependency and Exploitation in Bolivian Tin Mines*. New York: Columbia University Press, 1979.

New Idria Mining Company (Calif.) and William E. Barron. *Petition of the New Idria Mining Company to the President of the United States*. San Francisco: M. D. Carr & Co. Printers.

Noble, Bruce J., Jr., and Robert Spude, eds. *Guidelines for Identifying, Evaluating, and Registering Historic Mining Properties*. U.S. Department of the Interior, National Park Service, 1992.

Ord, Edward O. C. *Topographical Sketch of the Gold and Quicksilver District of California*. Philadelphia: P. S. Duval's Lith. Steam Press, 1848.

Palmer, Lyman. *History of Napa and Lake Counties*. San Francisco, California: Slocum, Bowen and Co. 1881.

Paul, Rodman W., and Elliott West. *Mining Frontiers of the Far West, 1848–1880*. Albuquerque: University of New Mexico Press, 2001.

Peachy, Archibald C., Gregory Yale, and United States District Court (California: Northern District). *The United States v. Andres Castillero: Mr. Peachy's Argument, on a Motion for a Commission to Mexico to Take Testimony*. San Francisco: s.n., 1860.

Pelanconi, Joseph Daniel. "Quicksilver Rush of Sonoma County, 1873–1875." M.A. thesis, Chico State College, California, 1969.

Pike, Ruth. *Enterprise and Adventure: The Genoese in Seville and the Opening of the New World*. Ithaca: Cornell University Press, 1966.

Pincetl, Stephanie S. *Transforming California: A Political History of Land Use and Development*. Baltimore: Johns Hopkins University Press, 1999.

Pitt, Leonard. *The Decline of the Californios*. Berkeley: University of California Press, 1966.

Pitti, Stephen. *The Devil in Silicon Valley*. Princeton, NJ: Princeton University Press, 2004.

Pitti, Stephen. "Quicksilver Community: Mexican Migrations and Politics in the Santa Clara Valley, 1800–1960." Ph.D. thesis, Stanford University, 1998.

Platt, Tristan. "Container Transport: From Skin Bags to Iron Flasks. Changing Technologies of Quicksilver Packaging between Almaden and America, 1788–1848." *Past & Present* 241, no. 1 (2012): 205–53. http://dx.doi.org/10.1093/pastj/gtr029.

Pred, Allan Richard. *Recognizing European Modernities: A Montage of the Present*. London and New York: Routledge, 1995.

Probert, Alan. "Bartolome de Medina: The Patio Process and the Sixteenth Century Silver Crisis." *Journal of the West* 3, no. 1 (1969): 90–124.

Quartz Miner. *Smart and Cornered: How to Get a Mine without Finding, Opening or Working One.* San Francisco: Commercial Book and Job Steam Printing Office, 1860.
"Quicksilver." *Mining & Scientific Press*, January 23, 1875, 60.
Quicksilver Mining Company. *Charter and By-Laws.* New York: H. Anstice & Co. Printers, 1883.
Quicksilver Mining Company. *Charter and Organization of the Quicksilver Mining Company, with Statement of Location, Extent of Property, &c.* New York: R. C. Root Anthony & Co., 1861.
Quicksilver Mining Company. *Reports and Exhibits Submitted at the Annual Meeting of the Stockholders.* New York: Sun Job Printing House, 1864.
Quicksilver Mining Company. *Reports and Exhibits Submitted at the Annual Meeting of the Stockholders.* New York: Sun Job Printing House, 1866.
Quicksilver Mining Company. *Reports and Exhibits Submitted at the Annual Meeting of the Stockholders.* New York: Sun Job Printing House, 1869.
Quicksilver Mining Company. *Statement of Earnings, Expenses, Profits and Dividends, for Twenty-One Years, Ending December 31st, 1891, under the Management of J.B. Randol.* N.p., 1892.
Ragsdale, Kenneth Baxter. *Quicksilver: Terlingua and the Chisos Mining Company.* College Station: Texas A&M University Press, 1976.
Randol, J. B. *California Quicksilver: Quicksilver Here and Abroad: Improvements in Quicksilver Mining: The Old and the New: A Comparison of the Two Noted Almadens: Their Wages and Products: No Quicksilver Monopoly.* San Francisco: s.n., 1890.
Randol, J. B. *Quicksilver and the Tariff in the House of Representatives and Elsewhere.* San Francisco: Randol, 1890.
Randol, J. B., and A. J. Leary. *California Quicksilver. Re-printed from the "Evening Bulletin," May, 1884. With Notes and Additions.* San Francisco: A. J. Leary Printer, 1884.
Randol, J. B., and United States Bureau of the Census. *Mines and Mining Quicksilver.* Washington, DC: s.n., 1890.
Randolph, Edmund, and Andrés Castillero. *District Court of the United States, Northern District of California. The United States vs. Andres Castillero. No. 366 U.S. Land Commission. No. 420 U.S. District Court. On Cross Appeal. Argument of Edmund Randolph, Assistant Counsel of the United States.* San Francisco: Towne & Bacon Printers, 1860.
Rawls, James J., and Walton Bean. *California: An Interpretive History.* Boston: McGraw Hill, 1998.
Rawls, James J., and Richard J. Orsi, eds. *A Golden State: Mining and Economic Development in Gold Rush California.* Berkeley: University of California Press, 1999.
Raymond, Rossiter W. *Mineral Resources of the States and Territories West of the Rocky Mountains.* Washington, DC: Government Printing Office, 1869.
Raymond, Rossiter W. *Mineral Resources of the States and Territories West of the Rocky Mountains.* Washington, DC: Government Printing Office, 1873.
Raymond, Rossiter W. *Statistics of Mines and Mining in the States*, 1873.

Raymond, Rossiter W. *Statistics of Mines and Mining in the States and Territories West of the Rocky Mountains for the Year 1870*. Washington, DC: Government Printing Office, 1872.

Revere, Joseph W., and Joseph A. Sullivan. *Naval Duty in California: With Map and Plates from Original Designs*. Oakland, CA: Biobooks, 1947.

Rickard, T. A. *Four Mining Engineers at New Almaden*. Ed. Gage McKinney. New Almaden, CA: New Almaden Quicksilver County Park Association, 1995.

Rickard, T. A. *A History of American Mining*. New York and London: McGraw-Hill Book Company, 1932.

Robbins, William G. *Colony and Empire: The Capitalist Transformation of the American West*. Lawrence: University Press of Kansas, 1994.

Robins, Nicholas. *Mercury, Mining, and Empire*. Bloomington: Indiana University Press, 2011.

Robinson, W. W. *Land in California*. New York: Arno Press, 1979.

Sassen, Saskia. *Globalization and Its Discontents: Essays on the New Mobility of People and Money*. New York: New Press, 1998.

Saxton, Alexander. *The Indispensable Enemy*. Berkeley: University of California Press, 1971.

Schneider, Jimmie. *Quicksilver: A Complete History of Santa Clara County's New Almaden Mine*. San Jose, CA: Zella Schneider, 1992.

Schuette, Curt N. "The Geology of Quicksilver Ore Deposits." *California Journal of Mines and Geology* (published by the State Division of Mines) 33, no. 1 (1937): 38–50.

Schuette, Curt N. *Quicksilver*. Bulletin 335. Washington, DC: Government Printing Office, 1931.

Schweikart, Larry, and Lynne Pierson Doti. "From Hard Money to Branch Banking: California Banking in the Gold-Rush Economy." In *A Golden State: Mining and Economic Development in Gold Rush California*, ed. James Orsi and Richard Rawls, 209–32. Berkeley: University of California Press, 1999.

Scranton, Philip. "Varieties of Paternalism: Industrial Structures and the Social Relations of Production in American Textiles." *American Quarterly* 36, no. 2 (1984): 235–57. http://dx.doi.org/10.2307/2712726.

Sewell, Jessica. "Gendering the Spaces of Modernity: Women and Public Space in San Francisco, 1890–1915." Ph.D. diss., University of California, Berkeley, 2000.

Sewell, Jessica. *Women and the Everyday City: Public Space in San Francisco, 1890–1915*. Minneapolis: University of Minnesota Press, 2011.

Sewell, William H., Jr. "A Theory of Structure: Duality, Agency, and Transformation." *American Journal of Sociology* 98, no. 1 (1992): 1–29. http://dx.doi.org/10.1086/229967.

Shinn, Charles Howard. *Mining Camps: A Study in American Frontier Government*. New York: Alfred A. Knopf, 1948.

Shutes, Milton H. *Abraham Lincoln and the New Almaden Mine*. San Francisco: L. R. Kennedy Printer, 1936.

Simpson, Brian. *Geological Maps*. Oxford: Pergamon Press, 1968.

Smedberg, James R. *Report of Trip to Quicksilver Mines, Dec. 16–27, 1879*. San Francisco, 1879.

Smith, Grant H., with new material by Joseph V. Tingley. *The History of the Comstock Lode, 1850–1997*. Reno: Nevada Bureau of Mines and Geology in Association with the University of Nevada Press, 1998.

Spence, Clark C. *British Investments and the American Mining Frontier, 1860–1901*. Moscow: University of Idaho Press, 1995.

Spude, Robert L. "Cyanide and the Flood of Gold: Some Colorado Beginnings of the Cyanide Process of Gold Extraction." In *Essays and Monographs in Colorado History*, no. 12, 1–35. Denver: Colorado Historical Society, 1991.

Spude, Robert L. "Mining History—A New Dialogue." *Cultural Resource Management* 7 (1998): 3.

St. Clair, David J. "New Almaden and California Quicksilver in the Pacific Rim Economy." *California History* (San Francisco) 73 (Winter 1994): 278–95.

Stegner, Wallace. *Angle of Repose*. New York: Penguin, 1971.

St. John, Lucy. "The Quicksilver Mine of New Almaden." *Our Young Folks: An Illustrated Magazine for Boys and Girls* [Boston: Ticknor & Fields] 7, no. 10 (October 1866): 590–597.

Sullivan, Frank J., and Charles N. Felton. *A Contested Election in California. __ vs. Hon. C. N. Felton. Testimony of the Qualified Electors and Legal Voters of New Almaden*. N.p., 1887.

Takaki, Ronald T. *Iron Cages: Race and Culture in 19th-Century America*. New York: Oxford University Press, 1979.

Takaki, Ronald T. *Strangers from a Different Shore: A History of Asian Americans*. New York: Penguin Books, 1990.

Thos. H. Thompson & Co. *Sonoma County Historical Atlas*. Oakland, CA, 1877.

Turner, Frederick Jackson. *The Frontier in American History: Mineola*. New York: Dover Publications, 1996.

Tyler, Ronnie C. *The Big Bend: A History of the Last Texas Frontier*. Washington, DC: Department of the Interior, National Park Service, 1975.

Udden, J. A. "The Anticlinal Theory as Applied to Some Quicksilver Deposits." *University of Texas Bulletin*, no. 1822 (1918): 7–30.

United States and John Parrott. *The United States of America v. John Parrott, Henry W. Halleck, James R. Bolton and Others in Equity*. 1858.

United States Congress, Joint Special Committee to Investigate Chinese Immigration. *Report of the Joint Special Committee to Investigate Chinese Immigration, February 27, 1877*. Washington, DC: Government Printing Office, 1877.

Upton, Dell. *Holy Things and Profane: Anglican Parish Churches in Colonial Virginia*. New Haven: Yale University Press, 1997.

U.S. Census Bureau. Manuscript Schedules of Decennial Population Censuses. 1860, 1870, 1880, 1900, 1910.

Valencia, Francisco, Antonio Soto, and Joe Graham. *Presentation on New Almaden Quicksilver Mines: Southwest Labor Studies Association Conference, San Jose State*

University, April 30, 1983. San Jose: Reproduced and distributed by the Chicano Library Resource Center, San Jose State University, 1998.

Wagoner, Luther. *Report upon the Guadalupe Quicksilver Mine: Geological Features.* San Francisco: s.n., 1881.

Walker, Richard. "California's Golden Road to Riches: Natural Resources and Regional Capitalism, 1848–1940." *Annals of the Association of American Geographers* 91, no. 1 (2001): 167–99. http://dx.doi.org/10.1111/0004-5608.00238.

Wallerstein, Immanuel. *The Modern World-System.* New York: Academic Press, 1974.

Wells, William. "How We Get Gold in California." *Harper's New Monthly Magazine* 20, no. 119 (April 1860): 605.

Wells, William. "A Visit to the Quicksilver Mines of New Almaden: Belonging to the Quicksilver Mining Company." *Harper's New Monthly Magazine* (June 1863), 25–40.

Wheat, Carl I. *Gold and Quicksilver District of California.* New York: James Gordon Bennett, 1848.

Whitaker, Arthur Preston. *The Huancavelica Mercury Mine: A Contribution to the History of the Bourbon Renaissance in the Spanish Empire.* Cambridge: Harvard University Press, 1941.

Wilkie, Laurie A. *Creating Freedom: Material Culture and African American Identity at Oakley Plantation, Louisiana, 1840–1950.* Baton Rouge: Louisiana State University Press, 2000.

Williams, Raymond. *The Country and the City.* New York: Oxford University Press, 1973.

Winks, Robin W. *Frederick Billings: A Life.* Berkeley: University of California Press, 1991.

Wolf, Eric R. *Europe and the People without History.* Berkeley: University of California Press, 1982.

Worster, Donald. *Under Western Skies: Nature and History in the American West.* New York: Oxford University Press, 1992.

Young, Otis E., Jr. *Western Mining: An Informal Account of Precious-Metals Prospecting, Placering, Lode Mining, and Milling on the American Frontier from Spanish Time to 1863.* Norman: University of Oklahoma Press, 1970.

Zanjani, Sally. *Goldfield.* Athens: Ohio University Press, 1992.

Zhu, Liping. *A Chinaman's Chance: The Chinese on the Rocky Mountain Mining Frontier.* Niwot: University Press of Colorado, 1997.

Zhu, Liping. "No Need to Rush: The Chinese, Placer Mining, and the Western Environment." *Montana, The Magazine of Western History* 49, no. 3 (1999): 42–57.

Index

Page numbers in italics indicate illustrations.

Abbott Mine, *111*, 112, 207, 212–13(n39)
Aetna Mine, 206, 211(n18)
Ah Cat, 203
Ah Shee, 203, 238
Almadén Mine, 3, 24, 37, 44, 45, 51(nn28, 30), 53(n47), 66, 71, 122; comparison to New Almaden, 34–35; Hapsburg control of, 39–40
Altoona Mine, 90(n76), 99, 133(n33), 211(n18), 212(n35)
amalgamation, mercury used in, 38, 42, 51(n25)
Americanization, of California, 16–17
American Town, 240
Anglo-Americans, 147, 186(n5), 193, 197, 198
Austrian Hapsburgs, 40
Avila, Patricio, *146*

Bañales, Juan, 139
Bancroft, Hubert Howe, 64
Bank Crowd, 68, 77, 90(n77), 94; Comstock production, 75–76, 79; and New Idria Mine, 93, 130(n2), 255; and Quicksilver Mining Company, 69, 88(n55)
banking houses, 38–39, 40
Bank of California, 79, 87(n37); and Quicksilver Mining Company, 68–69, 71, 72
Barron, Eustace, 27, 46, 48, 49, 53(n58), 83(nn1, 3), 259
Barron, William E., 57, 64, 68, 70, 83, 86(n23), 87(n35), 88–89(n57), 93, 199, 255
Barron & Co., 61, 62, 67, 68, 72, 86(n32), 88(n45), 133(n29); contracts, 70, 71, 75
Barron, Forbes & Co., 37, 46, 53(n58), 57, 83–84(n3), 86(nn23, 31), 87(n35), 139, 172, 190(n65), 245; and California, 48–49, 82–83; mercury sales, 58, 68; and New Almaden, 13, 14, 21, 22–23, 47–48, 54(n66), 55(n74), 60, 61, 62, 69, 147, 163, 178, 187(n21), 190(n65); Pacific market, 59, 60; trading network, 26–27
Barton, William H., 250
Becker, George F., 100, 101, 204
Bell, Thomas, 57–58, 64, 78, 83, 86(n23), 87(n38), 88–89(n57), 90–91(nn79, 84, 85), 93, 98, 245, 255; and Bank of California, 68–69; and Comstock production, 75–76; contracts and deals, 70, 80–81, 82; New Idria Mine, 130(n2), 199; North Bloomfield Mine, 74, 79, 253
Belmont, August, 67, 86(n31)
Big Bend (Tex.), *115*, *119*, *120*, 131(n14) 132(n17), 255
Billings, Frederick, 63–64
boardinghouses, *176*, 228, 233, 249
Bolivia, 45
Bolton, James Robert, 57, 83(n2)
Bolton, Barron & Co., 83(n3); mercury markets and, 58–60; operations of, 57–58
Bond, William, 66
bottling rooms, *167*, 168
Brass Wire Company, 161
Brewer, William, 51(n19), 61, 173, 243(n27); visit to New Almaden, 35–36
British Empire, 16, 82–83, 259; and Mexico, 46–47
Brown China camp, 238
Browne, J. Ross, 36, 51(n20)
Buchanan administration, 23, 65
Buckeye Mine, 212–13(n39), *258*
Buena Vista shaft, *158*, *160*
bullion production, 51(n26), 57, 66; Comstock, 75–76
Bulmore, Robert, 153, 187(n29); photos by, *152*, *154*, *155*, *157*–*58*
Burling, James, 247
Burling, William, 247
Burlingame Treaty, 251
Bustamante furnaces, *122*–23

277

Butterworth, Samuel, 51(n20), 63, 64–65, 72, 83(n3), 89(n66); at New Almaden, 66, 68, 69, 71, 74–75, 86(n24)

cabins. *See* houses
California, 1, 6, 21, 51(n24), 54(n69), 71, 114, 137; Americanization of, 16–17; Barron, Forbes & Co. and, 46–49, 82–83; Foreign Miners' Tax, 14–15; land ownership in, 60, 61–62, 186(n13); state constitution, 81, 246–47, 249–51
California: A History of Upper and Lower California from Their First Discovery to the Present Day (Forbes), 46–47
California Land Commission, 62
California Quicksilver Association, 62
Californios, 137, 139, 147, 172, 186(n5)
Calistoga, 128, 230, 231, 240, 249
Carson Mill Co., 77
Casa Grande, 15, 24, *25*, 35, 36, 50(n15), 51(n21), 65, 172, *173*, *174*
Castillero, Andrés, 24, 54(n66), 62, 83(n1), 85(n15)
Castillero Mine Claim, 63
Catholic church, Spanishtown, *175*, 178
Catholics, 196, 223
Central Americans, 15
chargers, 163, 165
Charles V, 40, 41
Chile, 35, 145–46
Chileans, 8, 172, 197, 199, 200; at New Almaden Mine, 13, 15, 32, 34, 139; racial and ethnic hierarchy, 98, 186(n5); tribute contracts, 145–46
China, 51(nn24, 28), 251; mercury trade in, 35, 53(n54), 59, 66, 71, 88(n47), 90(n79), 174
Chinese, 15, 81, 98, 128, 189(n51), 193, *197*, 210(n11), 211(n18), 213(n43); California's exclusion of, 210(n9), 246–47, 249–50, 253, 255; in company towns, 229–31, 242(n7); contract labor, 208, 218; at Great Western mine, 200–202, 203–4, 215, 216, *236–38*; at Guadalupe Mine, 198–99, 209–10(nn7, 10), 219; immigration, 251–52; in mining camps, 233–36, 240, 241, 244(n41); as ore pickers, 169–70, 195–96; in racial hierarchy, 138, 184, 186(n5), 217; as sojourners, 202–3; at Sulphur Bank Mine, 211–12(n29), 231–33, 247–49; supervision of, 204–5
Chinese pagoda, at Casa Grande, *173*, *174*

Christy, Samuel, 102
cinnabar, 11, 37, 134(n45); geology of, 100–104, 131–32(nn15, 16); prospects, 104–7
Cinnabar Mining District, 107, 109, 212(n35)
Civil War, American, 17, 61, 89(n65)
Clear Lake, 78, 103, 107, 230, *248*
Clear Lake District, 110
Coast Ranges, 27, 102, 128, 132(n18); mercury mines in, 6, *9*, *10*, 78, 255; mining districts in, 107, 109. *See also various mines by name*
colonialism, 16, 39, 45, 259–60
Colusa County, 110, 206
communities, mines as, 215–*16*
companies: Chinese, 203–4; miners', 150–51, 152–56, 197
company stores, 128, 129, 134–35(n51), 189(n51), 228, 241, 242(n7); New Almaden, *176*, *179*, 181–82; at Sulphur Bank Mine, 232–33
company towns: Chinese in, 229–31; establishment of, 180, 181–82; Guadalupe Village as, 218, 219–21; Knoxville as, 225–26; Sulphur Bank Mine, 232–33
Comstock, 75–76, 77, 79, 87(n37), 259
condensing systems, *120*, *166*, 167–68, 222
constitution, California state, 81, 246–47, 249–51
contract system, 44, 169, 171, 186(n6), 187(n21), 198, 210(n12); for aboveground work, 156–61, 188(nn36, 43); Chinese, 208, 218; at New Almaden, 32, 145–47, 150–51; for underground work, 152–56
Cornish, 98, *146*, 170, 183, 184, 186(n5), 188(n39), 199, 205, 209–10(n7); mining contracts, 147, 155–56, 161; at New Almaden, 139, 141, *152*, 153, 154, 179, 186(n7)
Corona Mine, *257*
cottages. *See* houses
Creoles, 45
cyanide process, 82, 260(n9)

Day, Sherman, *7*, 86(n24)
Disturnell, Nathaniel, 112, 206–7, 212(n38)
Disturnell, Richard O., 111, 112, 207, 212(n38)
Doige, William, *146*
Downer, Mrs. S. A., 7–8; description of New Almaden, 27–35, 139, 189(n48)
Drew, Daniel, 76, 89(n71)
drilling by the foot, 153

278 INDEX

Edwards Woodruff v. North Bloomfield Mining and Gravel Company, 245, 252–53
Eldridge, James, 65, 86(n24)
Emmert, Paul, illustration by, 23
Empire Claim, 207
English Americans, 170, 186(n5)
Englishtown, *149, 176, 177,* 178–*79,* 180, 185
environmental damage, 255, 259
ethnicity, 98, 130, 137–38, 139, 190(n58), 193, 208, 242(nn4, 5); contract system, 151–56; foreign miners' tax, 14–15; landscapes of, 18–19. *See also* race
Euro-Americans, 209–10(nn7, 10); at Guadalupe Mine, 198, 199, 218. *See also* whites

Fiedler, Superintendent, 250
Foote, Mary Hallock, 190(n62), 241; on Hacienda landscape, 173–74
Forbes, Alexander, 27, 48, 49, 53(n58), 54(nn64, 66), 83(nn1, 3), 86(n23), 259; on British control of California, 46–47
Forbes Quicksilver Mine, *5*
Foreign Miners' Tax, 14–15
Fossatt case, 63, 64
Fuggers, 38–39, 40, 51(n31), 53(n56)
furnaces, 131(n6), 210(n11), *222,* 260(n4); Bustamante, *122–23*; continuous-feed, 125–27, 134(n49), 189(n56), 211(n19), *257, 258*; Idria/Idrija, 123–*124*; at New Almaden, 161–67, 188–89(nn44, 45, 46); retort, 51(n37), 112, *113*
furnace yards, contractors in, 160–67, 188(n43)

Garnes, Jose, 183
geology, 110, 132(n18), 133(n31), 190(n57); cinnabar, 100–104, 131–32(nn15, 16), 201; New Idria Mine, 94, *95*
Germans, 193
Getz Brothers Company, 232–33, 249
Geysers, 107, 206, 240
Goodwin, A. C., 107
gold mines/mining, 1, 12–13, 38, 39, 41; hydraulic, 74, 81–82; locations of, *9,* 11
Goss, Helen Rocca, 233–35
Great Britain, 13; mercantile system, 55(n74), 82; and Mexico, 46–47, 48–49
Great Eastern Mine, 79, 90(n76), 132–33(n24), 211(n18), 247

Great Western Mine, 51(n24), 79, 81, 89(n68), 99, 110, 132–33(n24), 134(n49), 196, 209, 230, *232,* 241, 242(n3), 249, 253; Chinese workers at, 128, 203–4, 233–38, 247, 255; community at, 215–*16*; organization of work at, 200–202
Guadalupe District, 110
Guadalupe Mine, 80, 96, 98, 134(n49), 146, 194, 209–10(n7), 213(n43), 217, 233, 242(n11); expenditures, 129–30, 134–35(n51); mining camp, 127, 218; organization of work at, 197, 198–99, 208–9, 210(n9)
Guadalupe Mining Company, 198
Guadalupe Village, *226,* 228, 242(n12); houses at, 219–21, *223, 224*; structure of, 218–19, 222, *225,* 242(n12)
guilds, miners', 145–47

Hacienda Nuevo Almaden, 15, *28, 29, 34, 162, 171,* 182, 185, 190(n66), 259; Barron, Forbes & Co. and, 47–48, 49; landscape of, 24–27, 173–74, 176; reduction works at, 149, 150; structure of, 172–73
hacienda system, 26, 50(n110), 66, 98, 217
Halleck, Henry Wager, 36, 63–64, 84(n12)
Halleck, Peachy & Billings, 36
Hapsburgs, 39–40, 51(n29)
Harry, James, *152, 154,* 187(n26)
Helping Hand Club, *183,* 185
Hermoza & Co., 66–67
Hindus, at New Idria, 193
Hoffman, Judge, 251
Holy Roman Empire, 40
Hong Kong, mercury shipments to, 59, 71
Hood, James, 198
houses/housing, 190(n59), 243(nn14, 24), *256*; at Englishtown, *177,* 178, 179; at Great Western Mine, 234–35; at Guadalupe Village, 218, 219–21, *223, 224*; at Spanishtown, *175,* 177–78; at Sulphur Bank Mine, 232–33
Huancavelica Mine, 37, 43–44, 45, 53(n53)
Hüttner-Scott furnace. *See* Scott furnace
hydraulic mining, 74, 81–82, 253

Idria (Calif.), sale of, *254,* 255. *See also* New Idria Mine
Idria/Idrija furnaces, 123–*124*
Idrija Mine (Slovenia), 3, 24, 37, 40, 41, 42, 44, 45, 51(nn23, 30), 53(nn47, 51)

INDEX 279

Indians, 139; in racial hierarchy, 137, 138
Integral Mine, 99, 212(n35)
Irish, 98, 160, 170, 213(n43), 243(n17); at Redington Mine, 196, 199, 200, 208, 209–10(n7), 223, 225
Italians, 193

Jennings, Superintendent, 153, 170

Kearney, Denis, 246
Klau Mine, 99
Knox, Richard, 124–25, 134(n49)
Knox-Osborn furnaces, *125–27*, 134(n49), 260(n4)
Knoxville, 200, 210–11(nn14, 15), 227; as company town, 98–99, 225–26; structure of, 223–24, 243(n17)
Knoxville District, 110
Koreans, 193

labor, 32, 53(n47), 98, 139, 188–89(n45); Chinese, 203, 253; contract, 150, 161; costs of, 228–29; day and monthly, 151, 188(n34); forced, 43–44; organization of, 196, 204–5
Lake County, *248*; mercury mines in, *10*, 11, 90(n76), 99; mining districts, 107, 110, 133(n32), 206
land claims, 36, 186(n13)
landownership, 85(n13); in California, 60, 61–62, 186(n13); mining claims, 13–14
landscapes: Guadalupe Village, 219–23, *225*; Hacienda Nuevo Almaden as, 173–74, 176; New Almaden Mine, 36, *149*–50; mine and camp, 112–13, 216–17; paternalism and, 180–81; race and ethnicity, 18–19, 137–38; of work and life, 17–18
Larios Grant, 64, 85(n15)
Larkin, Thomas, 48, 50(n8); as U.S. consul, 21–24
Lightner, H. S., 247
Lincoln administration, 23, 63, 64
Lower Lake, 230

Maestrazgos, 40, 51(n31)
Magic Quicksilver Ring, 67, 70, 75, 76–77, 78–79, 99
Malakoff Diggings, *74*
Manhattan Mine, Knox-Osborn furnace in, 124–25

Manzanita Mine, *111*
Maria Christina, Queen Regent, 45
Mariposa County, 106
Mariscal Mine, *115*, *119*, *120*, 131(n14)
markets, 65, 71, 72; Pacific, 58–60; price drops, 79–80; Quicksilver Mining Company and, 67–68
Mayacamas District, 107, *108*, 109, 110, 206
MEB. *See* Merchant's Exchange Bank
Medina, Bartolomé de, patio process, 41, 42
Merchant's Exchange Bank (MEB), 80–81, 90(n77)
mercury: supplies of, 42–43, 45; uses of, 38, 51(n25), 260(n13)
mercury poisoning, 94, 168, 169, 189(nn48, 49, 50), 210–11(n15), 212(n31)
mercury production, 6, 13, 16, 52(n41), *73*, 84(n8), 88(n50), *129*; control of, 70–71; knowledge of, 2–3; monopoly, 38–39, 48; profits and, 76–77
mercury trade, 16, 35, 66, 84(n8), 89(n74); control of, 53(n52), 69–70; monopoly, 38–39; Pacific markets, 58–59; Rothschild control of, 45–46, 67, 68
Methodist Episcopal church, at New Almaden, *176*, 178
Mexican Method mining, 7, *141*
Mexicans, 8, 15, 160, 193, 208, 209–10(nn7, 10), 213(n43), 240, 255; and company stores, 181–82; at Guadalupe Mine, 198, 199; at New Almaden Mine, 13, 32, 34, 139, 169, 170; racial and ethnic hierarchy, 98, 137, 147, 186(n5); at Redington Mine, 196, *197*, 200, 226, 229; in Spanishtown, 172, 180; strikes, 183, 184, 191(nn74, 75); as trammers, *155*, 163; tribute contracts, 145–46, 153, 154, 156, 186(n7)
Mexico, 45, 51(n24), 54–55(n70), 59, 63, 186(n7); British Empire and, 46–47, 48–49; mercury trade, 66–67; mercury use, 38, 88(n47); silver rush, 41, 51(n34), 52(n40); tribute system, 145–46
Middletown, 128, 230, 231
Mills, D. O., 68, 72, 79, 87(n37), 130(n2)
Mine Hill, 24, *25*, 29–30, *33*, 96, 149, *162*, 177, 190(n57)
mine managers, 150–51; on strikes, 183–84
miners, 81, 98, 138–39; guilds and, 145–47; New Almaden, 15, *30*, 32, 34, 149, 150; New Idria, 193, 194. *See also various jobs; tasks*

Miners' Association of California, 253
miners' convention, San Carlos District, 106
mines, 9–11, 132(n23); economy of, 128–30; environmental damage of, 255, 259; goals of, 194–95; hardrock, 138–39; landscape of, 112–13; physical and social contexts of, 94, 96, 98; structure of, 114–17, 132(n19), 134(n42). *See also by name*
mining: aboveground work, 156–69; capital for, 12–13; contract, 150–51; systems of, 54(n66), 139–40; underground work, 152–56
mining camps, 127–28, 240–41; under Barron, Forbes & Co., 180–81; Chinese at, 229–38; landscapes of, 216–17; under Quicksilver Mining Company, 181–84; racialized, 171, 172–80; structure of, 112–14
mining claims, 13–14, 213(n40)
mining districts, 14, 99; boom, 238–39, 241; promotion of, 239–40; quicksilver, 106, 107–12, 212(n35); work structures in, 205–7
mining engineers, 114
mining grants, Mexican, 24
mita system, 43–44, 52–53(n46)
Mock Bing Yer, *249*
monopoly, mercury production and trade, 38–39, 60, 75, 77, 79
Monterey, 98
Monterey County, 106
Mormon Gulch, quicksilver machine, *12*

Napa Consolidated Mine (Oat Hill Mine), 6, 51(n24), 79, 81, 99, 110, 201, 204, 253; Chinese camp at, 233, *235*; Chinese labor at, 247, 250, 252, 255
Napa County, 78, 211(n18), 224, *248*; mercury mines in, *10*, 11, 70, 99; mining districts, 107, 133(n32)
Napoleon Bonaparte, 44, 45
New Almaden, 15, 50(n15), 127; description of, 27–28; landscape of, 24–27, 137, *149*
New Almaden Company, 13, 14, 15, 24, 36, 63, 85(n15); U.S. interests in, 63–64, 87(n41)
New Almaden District, 107, 109, 110
New Almaden Mine, 2, 3, 13, 17, 37, 54(n64), 70, 78, 81, 83(n1), 86(n32), 96, 104, 107, 130(n2), 132(n19), 134(n45), 190(n66), 194, 208–9, 210(n8), 241, 259; aboveground work at, 156–69, 188(n36); Barron, Forbes & Co. management of, 47–48; Chinese at, 195–96, 253, 255; contract system, 146, 161, 187(n21); furnaces, *123*, *126*, 134(n49), 188–89(nn44, 45, 46); infrastructure of, 139–45; landscape, 36, 149–50; Thomas Larkin's report on, 21–24; location of, *10*, 11; managers of, 150–51; marketing and promotion of, 65–66; mercury production, 70–71, *73*, 90(n81); mining camps associated with, 127–28, 172–80; ore grades at, *121*–22, 188(n37); ownership of, 55(n74), 61–63, 64, 83(nn2, 3); Pacific market, 58–60; production, 50(n8), 54(n67), 72, 77, 88(n50), 209; Quicksilver Mining Company and, 180–82, 183–84; racial hierarchy of, 147–48, 169–71, 186(n5), 199, 217; James B. Randol and, 74–75, 89(n66), 184–85; strike at, 182–83, 191(n74); technology, 114, 117, 122, *123*, *126*; underground work at, *152*–56; visitors' accounts of, 6–8, 27–36
New Idria Company, 255, 260(n12)
New Idria District, 106, 109, 110
New Idria Mine, 6, 19(n11), 50(n10), 51(n24), 88(n44), 93, 96, *97*, 130(n2), 131(n6), 146, 210(n12), 217, 243(n27), 253, *256*; Bell's control of, 80–81; Chinese labor at, 170, 210(n11); geology of 94, *95*, 104; location of, *10*, 11; mercury production, 70–71, 72, *73*, 75, 76, 78–79, 88(n50), 90(n81); organization of work at, 197, 199, 209(n2); ownership of, 60, 61, 64, 83(n3); sale of, *254*, 255; square set mines in, 117, *118*; workers at, 193, *194*, *195*
North Bloomfield Gravel and Mining Company, 72, 74, 88(n56); injunction against, 81–82; lawsuit against, 252–53

Oat Hill Mine. *See* Napa Consolidated Mine
O'Brien, William, 89(n70), 198
Oceanic Mine, 90(n76), 99, 133(n33), 211(n18)
Ophir Mine, 79
ore delivery systems, *119*
ore pickers, Chinese as, 169–70, 195–96
ore processing, 139, 171; contract system, 151, 156–57, 188(nn37, 39); at furnaces, 161–67; sorting, 157–61
Osborn, Joseph, 125, 134(n49)

Parrott, John, 83(n2), 90(n76), 246, 247, 252
Parrott, Tiburcio, 80, 81, 132–33(n24), 247, 250–51; court case, 245–46, 252

Parrott & Company, 83(n2), 204, 242(n3), 252, 260(n4)
paternalism, 190–91(n67), 217–18; Barron, Forbes & Co., 180–81; Quicksilver Mining Company, 181–84
patio process, 38; invention of, 41–42; mercury use in, 42–43, 44; at New Almaden, 32, *33*
peddlers, Mexican, 182
Peru, 41, 43, 45, 52(n40)
Pine Flat area, *108*
Pine Flat Quicksilver District, 11, 14, 99, 107, 132(n23), 206, 211(n18), 212–13(nn39, 40), 238; tourism, 239–40
Pioneer; or, California Monthly Magazine, The, account of New Almaden Mine, 27–35
placer mining, 12–13
planillas, 198; contract work at, 156–60; as outmoded, 170–71
pollution, 259
Potosí, silver mines at, 41, 44
Progressive Era reforms, 183, 185
prospects, prospecting, 14; cinnabar, 78, 104–7, 206–7
Protestants, 196, 223, 243(n29)

QMC. *See* Quicksilver Mining Company
quicksilver booms, 238, 241; organization of work and, 200–207, 208–9, 210(n8); promotion of, 239–40
Quicksilver District, 3–5
Quicksilver Machine, *12*
Quicksilver Mining Company (QMC), 71, 86(nn24, 31), 88(n55), 93, 171, 178; and Bank of California, 68–69; contracts, 70, 75; New Almaden Mine ownership, 61, 62–63, 64, 65–66, 87(n43); paternalism of, 180–84; racialized work structure of, 148–49; strikes against, 182–84; world markets and, 67–68

race; racial hierarchy, 98, 130, 137–38, 139, 186(n5); aboveground work, 160, 161; Chinese in, 217–18; contract system, 151–52; foreign miners' tax, 14–15; landscapes of, 18–19; miners and, 147–48; at mining camps, 172–80, 242(n12); at New Almaden, 169–71; at New Idria, 198–99; organization of work and, 204–5, 207–8; Quicksilver Mining Company, 148–49, 180–84; at Redington Mine, 199–200; underground work, *152–56*
racialization, 137; mining camp, 216–17; of New Almaden, 147–48, 176–80
railroads, 90(n84), 128, *129*
Ralston, William C., 68, 69, 74, 75, 79, 87(n37), 130(n2)
Randol, James B., 80, 133(n25); at New Almaden, 74–75, 76, 77, 89(n66), 161, 184–85
Randol planilla (New Almaden), *157–59*
Randol shaft (New Almaden), *143*, *144*, *160*, 171
Raymond, Rossiter W., 71
Redington, John, 70, 210(n13), 223
Redington Company, 72
Redington Quicksilver Mine, 78, 80, 88(n45), 89(n70), 110, 127, 133(nn25, 32), 189(n5), 194, 201, 210–11(nn15, 16), 213(n43), 216, 217, 241, 253; ethnicity of workers at, 196, *197*; furnace technology in, 123, *124*; and Knoxville, 98–99, 223–27; mercury production, 70–71, *73*, 88(n50); organization of work at, 199–200, 208, 209–10(n7), 224–25; profits from, 71–72, 228–29; work stability in, 227–28
reduction works, 50(n10), 114, 139, 198; at New Almaden, 22, 23, 25, 32, *34*, 149, 150, 161–63; at New Idria, 94, 131(n6), *256*; technology of, 117, 119–27, 211(n19)
reduction yard, at New Idria, 193, *194*
retort furnaces, 51(n37), 112, *113*
Rocca, Andrew, 128, 203–4, 215, *216*, 238; on Chinese workers, 236, 237, 249–50
Rocca, Florence, 235–36
Romans, 38
Rothschild, Alphonse, 45
Rothschild cartel, 2, 38–39, 69, 77, 82, 86(n27), 259; and Bolton, Barron & Co., 58, 59; mercury trade, 45–46, 53(n56), 67, 68, 86(nn31, 32)

Sacramento Valley, 253
Saint John's Mine, 211(n18), 250
salivation, 94, 168, 169, 189(nn48–50), 210–11(n15)
Sampson, E. B., 93, 169
San Carlos District, 106
San Carlos Mine, *97*
San Francisco, 60; as supply center, 130, 134–35(n51)
San Francisco Mining Exchange, 79
San Jose, 98, 130, *158*, 191(n71)

San Luis Obispo County, *10*; mines in, 84(n10), 86(n76), 99, 211(n18)
San Luis Obispo District, 110
Santa Barbara County, 105
Santa Clara Mining Company, 198
Santa Ynez Valley, 105
Sawyer, Lorenzo, 81, 251, 253
Sawyer Decision, 253
Schneider, Jimmie, 76
Schuette, Curt N., 102, 115, 131–32(nn14, 15)
Scott furnace, *126*, 127, 134(n49), 166, *257*, *258*
serpentine, 104
Sharon, William, 79, 87(n37), 90–91(nn77, 85)
Sierra Nevada, gold mines in, *9*, 11
silver, silver mining, 39, 45, 51(nn33, 34); New World, 40–41; patio process, 38, 41–42; at Potosí, 43, 44
Sistema del Rato, El, 139, *140*, *141*, *142*
Six Companies, 203, 204
skip fillers, 155, 156
slagmen, 163, 166–67
social hierarchy, 98
Sonoma County, 78, 103, 211(n18); mercury mines in, *10*, 11, 90(n76), 99; mining district in, 107, 206
sorting floors. *See* planillas
South America, 48, 88(n47)
South Americans, 15. *See also* Chileans
Spanish Empire, 16, 45, 53(n50), 82, 259; mercury supplies and, 43–44; mercury trade and, 38–40, 52(n41), 53(n52); New World mines, 40–41, 52(n40)
Spanish Hapsburgs, and Almadén Mine, 39–40, 42
Spanish speakers *146*, 193, 218; at New Almaden Mine, 148, 172, 182–84. *See also* Californios; Chileans; Mexicans
Spanishtown, 15, *31*, *149*, *157*, 172, *175*, 180, 181, 190(n65); strikes staged at, 182–83; structure of, 176–79
square sets, in mining, 117, *118*
Stevenson, Robert Louis, 240
stock market, Comstock, 79
stope timbering, *145*
strikes, at New Almaden mine, 148, 182–84, 191(nn74, 75)
Sulphur Bank Mine, 79, 81, 89(n68), 90(n76), 99, 107, 110, 209, 212(n31), 230–31, 241, 250, 251, 253, 260(n3); Chinese at, 204, 211–12(n29), 231–33, 247–49, *252*; operations of, 194, 201
Sulphur Bank Quicksilver Mining Company, 247
Sulphur Creek Mining District, 99, 107, 110, *111*, 112, 206–7, 238, *258*
Sunderland, Thomas, 69
Swedes, 155, 156, 170
Swetl, Leonard, 63, 85(n19)
Swiss, 98

tariffs, U.S., 77, 89(n65)
taxation, of "foreign" miners, 14–15
teamsters, 160
technology, 18, 114–17, *118*; at New Almaden, 139–45; and organization of work, 208–9; at reduction works, 119–27, 131(n6), 211(n19)
tenateros, 142
Texas, 51(n24), *115*, *119*, *120*, 131(n14), 132(n17), 255
Thompson, Granville, 239
Thompson, Greenville, 239
Tiburcio Parrott, In re, 245–47, 250–52
Toledo, Francisco de, 43, 44, 53(n50)
tourism, in Pine Flat District, 239–40
towns, as supply centers, 128, 129, 130, 134–35(n51)
trade: international, 26–27, 35; Pacific, 58–60, 61
trammers, 154, *155*, 156, 163, 199
Treaty of Guadalupe Hidalgo, 84(n12), 147
tribute contracts/workers, 145–46, 153, *154*, 156, 161, 171
Trinity County, 90(n76), 99, 211(n18), 212(n35)
Trinity Group, 110

unions, at Redington Mine, 98
United States: and California, 21, 48, 54(n64); and New Almaden Mine, 23, 63–64
U.S. Army, 63
U.S. Supreme Court, and New Almaden ownership, 23, 60, 63, 64, 87(n41)

Venice, 40
villages. *See* mining camps; *by name*
Virginia and Truckee Railroad, 90(n84)
visitors, to New Almaden Mine, 6–8, 27–35

wages, 151, 153, 169, 210(n10), 213(n43), 227
whites, 196, 213(n43); at Great Western Mine, 215, *216*; in racial hierarchy, 137, 160, 169, 217; work hierarchy, 204–5

Wilkersheim, Fritz, illustration by, *22*
Wong Ah Sing, 246
Woodruff, Edwards, 253
work contracts, 145–47, *197*
workers, 138–39; aboveground, 156–69; at New Almaden, 148–50, 187(n24); racialization of, 147–48, 207–8; at Redington Mine, 227–28; underground, *152–56*

Workingman's Party, 246, *252*

yardage contracts/workers, *152*, 153, 161, 171, 186(n7), 193, *194*, 209(n2)
Yer Hop & Co., 249

Zellerbach, Anthony, 80, 81, 90(n76), 110, 133(n33)

www.ingramcontent.com/pod-product-compliance
Lightning Source LLC
Chambersburg PA
CBHW071151070526
44584CB00019B/2745